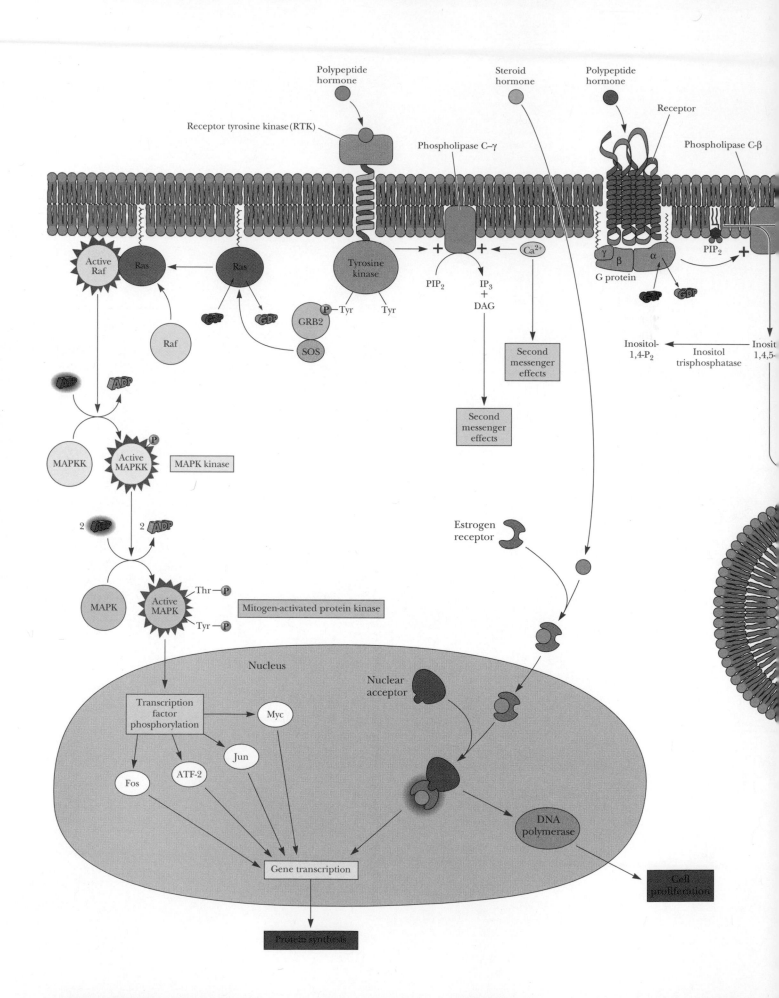

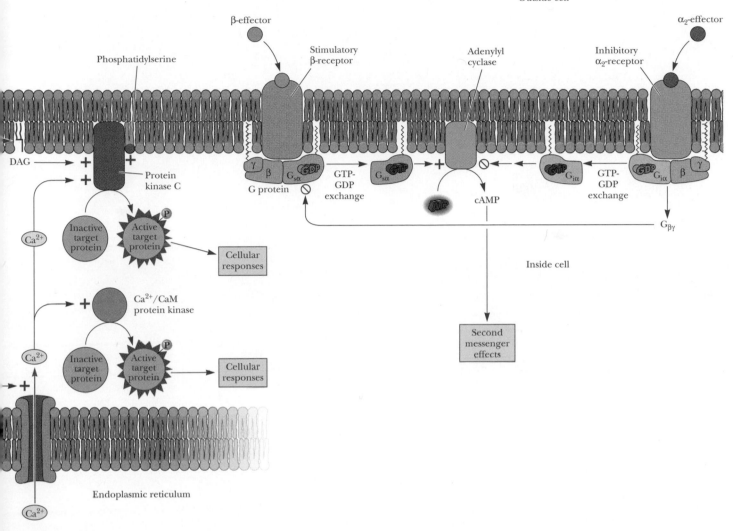

SIGNAL TRANSDUCTION PATHWAYS

Molecular Aspects of Cell Biology

Molecular Aspects of Cell Biology

Reginald H. Garrett

Charles M. Grisham

University of Virginia

SAUNDERS COLLEGE PUBLISHING

HARCOURT BRACE COLLEGE PUBLISHERS

Fort Worth • Philadelphia • San Diego • New York • Orlando • Austin

San Antonio • Toronto • Montreal • London • Sydney • Tokyo

Text Typeface: Baskerville
Compositor: York Graphic Services, Inc.
Acquisitions Editor: John J. Vondeling
Developmental Editor: Sandra Kiselica
Managing Editor: Carol Field
Project Editors: Becca Gruliow, Sarah Fitz-Hugh
Copy Editor: Zanae Rodrigo
Manager of Art and Design: Carol Bleistine
Art Director: Carol Bleistine, Anne Muldrow
Art Assistant: Sue Kinney
Text Designer: Rebecca Lemna
Cover Designer: Lawrence R. Didona
Text Artwork: J/B Woolsey Associates
Layout Artwork: Claudia Durrell
Director of EDP: Tim Frelick
Production Manager: Charlene Squibb
Marketing Manager: Marjorie Waldron
Field Product Manager: Laura Coaty

Cover Credit: J/B Woolsey Associates

Printed in the United States of America

4 5 6 7 8 9 0 1 2 3   032   10 9 8 7 6 5 4 3 2 1

ISBN   0-03-007597-1

Library of Congress Catalog Card Number: 93-087782

We dedicate this book to every student of biochemistry and to the professors who taught us biochemistry, especially

Alvin Nason	*Kenneth R. Schug*
William D. McElroy	*Ronald E. Barnett*
Maurice J. Bessman	*Albert S. Mildvan*
Ludwig Brand	*Rufus Lumry*

Preface

. . . .

Scientific understanding of the molecular nature of life is growing at an astounding rate. Significantly, society is the prime beneficiary of this increased understanding. Cures for diseases, better public health, remediations for environmental pollution, and the development of cheaper and safer natural products are just a few practical results of this knowledge.

In addition, this expansion of information furthers what Thomas Jefferson called "*the illimitable freedom of the human mind.*" Scientists can use the tools of biochemistry and molecular biology to explore all aspects of an organism—from basic questions about its chemical composition, through inquiries into the complexities of its metabolism, its differentiation and development, to analysis of its evolution and even its behavior. Biochemistry is a science whose boundaries now encompass all aspects of biology, from molecules to cells, to organisms, and to ecology.

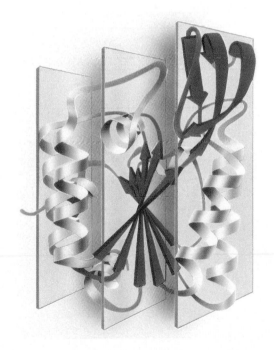

As biochemistry increases in prominence among the natural sciences, its inclusion in undergraduate and beginning-graduate curricula in biology and chemistry becomes imperative. And the challenge to authors and instructors is a formidable one: how to provide a comprehensive description of biochemistry in an introductory course or textbook. Fortunately, the increased scope of knowledge allows scientists to make generalizations connecting the biochemical properties of living systems with the character of their constituent molecules. As a consequence, these generalizations, validated by repetitive examples, emerge in time as principles of biochemistry, principles that are useful in discerning and describing new relationships between diverse biomolecular functions and in predicting the mechanisms underlying newly discovered biomolecular phenomena.

We are both biochemists—one of us is in a biology department, and the other is in a chemistry department. Undoubtedly, our approaches to biochemistry are influenced by the academic perspectives of our respective disciplines. We believe, however, that our collaboration on this textbook represents a melding of our perspectives that will provide new dimensions of appreciation and understanding for all students.

FEATURES AND ORGANIZATION

The organizational approach we have taken in this textbook is traditional in that it builds from the simple to the complex. The main body of the parent textbook, published separately as *Biochemistry,* is organized in four parts:

Part I, *The Molecular Components of Cells,* Part II, *Enzymes and Energetics,* Part III, *Metabolism and Its Regulation,* and Part IV, *Genetic Information.* This supplemental volume, titled *Molecular Aspects of Cell Biology,* describes the biochemistry that underlies a number of complex biological structures and systems. Its subject matter extends the range of topics that can be addressed in a contemporary biochemistry course. Together with the parent volume, this supplement provides pedagogical resources suitable for comprehensive two- or three-semester courses in biochemistry and molecular cell biology. In addition, *Molecular Aspects of Cell Biology* provides a more thorough coverage of selected molecular aspects of cell biology than is typically presented in texts on the molecular biology of the cell.

It begins with a discussion of the spontaneous assembly of large biomolecular aggregates. Chapter 34 examines the thermodynamic basis for spontaneous self-assembly and then describes several typical self-assembling systems, including microtubules and viruses. It concludes with a detailed examination of the structure and assembly of HIV, the AIDS virus. (Other classic self-assembling systems such as ribosomes are described in detail elsewhere in the text.) Chapter 35 explains biological transport processes, and Chapter 36 discusses the structure and function of muscle. Chapter 35 is notable for its coverage of recent developments related to several novel transport systems (including the multidrug transporter and the osteoclast proton pumps of bone) and amphipathic, ion-translocating peptides. Part V concludes with chapters on hormonal signaling (Chapter 37) and neurotransmission (Chapter 38). These chapters provide a fresh and up-to-date perspective on the rapidly changing fields of cellular signaling and signal transduction, with coverage of the superfamilies of membrane receptors, oncogenes, tumor suppressor genes, sensory transduction, and the biochemistry of neurological disorders.

Features

The newly designed and consistently colored illustrations include special icons for specific molecules, such as ATP, NADH, and coenzyme-A.

Computer-generated molecular structures, developed at the University of Virginia for this text, show the close relationship between molecular structure and biological function.

Critical Developments in Biochemistry boxes pique students' interest with information on new advances and technologies.

A Deeper Look boxes closely explore special-interest topics, such as the metabolic peculiarities of exotic plants and animals.

Interesting end-of-chapter problems and up-to-date references test students' mastery of the material and encourage further exploration.

Evocative chapter-opening photographs, coupled with literary or whimsical quotations, stimulate the imagination and serve as metaphors for chapter topics.

Acknowledgments

We are indebted to our colleagues David Jemiolo (Department of Biology, Vassar College) and Barton K. Hawkins (Department of Chemistry and Biochemistry, University of South Carolina), who carefully reviewed the entire manuscript at several stages, and to all the many other outstanding and invaluable expert reviewers of the manuscript.

We wish to warmly and gratefully acknowledge many other people who assisted and encouraged us in this endeavor. These include the outstanding staff at Saunders College Publishing, including our publisher, John Vondeling, who recruited us to this monumental task, and also Sandi Kiselica, Sarah Fitz-Hugh, Becca Gruliow, Carol Bleistine, Anne Muldrow, Carol Field, Tim Frelick, and Pauline Mula. The beautiful illustrations that grace this textbook are a testament to the creative and tasteful work of John and Bette Woolsey, Patrick Lane, and the entire staff of J/B Woolsey Associates. We are grateful to our many colleagues who provided original art and graphic images for this work, particularly Professor Jane Richardson of Duke University, who provided numerous original line drawings of the protein ribbon structures, and Mindy Whaley, who prepared the molecular graphic displays. We owe a very special thank-you to our devoted wives, Catherine Leigh Touchton and Rosemary Jurbala Grisham, to our children, Jeffrey, Randal, and Robert Garrett, and David, Emily, and Andrew Grisham, and finally to Cassie the cat, who stayed close to the project throughout its many stages. With the publication of this book we celebrate and commemorate the lives of our parents, William W. Garrett and Lelia B. Bosley, and Ernest M. Grisham and Mary Charlotte Markell Grisham.

Reginald H. Garrett   **Charles M. Grisham**
Advance Mills, VA   **Ivy, VA**
                                 February 1994

Support Package

The **Student Study Guide** by David Jemiolo (Vassar College) includes summaries of the chapters, detailed solutions to all end-of-chapter problems, a guide to the key points of each chapter, important definitions, and illustrations of major metabolic pathways.

The **Test Bank** by William Scovell (Bowling Green State University) includes over 900 multiple-choice questions for professors to use as tests, quizzes, and homework assignments. The bank of questions is also available in computerized form for IBM-compatible and Macintosh computers.

A set of **Overhead Transparency Acetates** includes 175 of the pedagogically most important figures in the text.

Monthly updates of **UVa Images,** a collection of macromolecular structures with explanatory text is available on Internet by accessing the computer facility at the University of Virginia. Users have the ability to manipulate the images in any dimension.

Saunders Chemistry Videodisc Version 3 Multimedia Package includes still images from the text, as well as hundreds from other Saunders chemistry texts. The disc can be operated via a computer, a bar code reader, or a hand-controlled keypad. It also features molecular simulations and demonstrations.

Lecture Active™ Software enables instructors to customize their lectures with the Videodisc. Available for both IBM and Macintosh computers.

List of Reviewers

Lutz Birnbaumer
Baylor College of Medicine

Michael Brown
Emory School of Medicine

Scott W. Champney
East Tennessee State University, College of Medicine

David J. DeRosier
Brandeis University

J. Martyn Gunn
Texas A&M University

Robert J. Kadner
University of Virginia, School of Medicine

Sidney Kushner
University of Georgia

Weng-Hsiung Li
University of Texas at Houston

Ponzy Lu
University of Pennsylvania

Neil Osheroff
Vanderbilt University, School of Medicine

Andrew Somlyo
University of Virginia, School of Medicine

Alfred Stracher
State University of New York, Health Science Center

Peter J. Wejksnora
University of Wisconsin, Milwaukee

About the Authors

· ·

Reginald H. Garrett was educated in the Baltimore city public schools and at the Johns Hopkins University, where he received his Ph.D. in biology in 1968. Since that time, he has been at the University of Virginia, where he is currently professor of biology. He is the author of numerous papers and review articles on biochemical, genetic, and molecular biological aspects of inorganic nitrogen metabolism. Since 1964, his research interests have centered on the pathway of nitrate assimilation in filamentous fungi. His investigations have contributed substantially to our understanding of the enzymology, genetics, and regulation of this major pathway of biological nitrogen acquisition. His research has been supported by grants from the National Institutes of Health, the National Science Foundation, and private industry. He is a former Fulbright Scholar and has been a Visiting Scholar at the University of Cambridge on two sabbatical occasions. He has taught biochemistry at the University of Virginia for 26 years. He is a member of the American Society for Biochemistry and Molecular Biology.

    Charles M. Grisham was born and raised in Minneapolis, Minnesota, and educated at Benilde High School. He received his B.S. in chemistry from the Illinois Institute of Technology in 1969 and his Ph.D. in chemistry from the University of Minnesota in 1973. Following a postdoctoral appointment at the Institute for Cancer Research in Philadelphia, he joined the faculty of the University of Virginia, where he is professor of chemistry. He has authored numerous papers and review articles on active transport of sodium, potassium, and calcium in mammalian systems, on protein kinase C, and on the applications of NMR and EPR spectroscopy to the study of biological systems. His work has been supported by the National Institutes of Health, the National Science Foundation, the Muscular Dystrophy Association of America, the Research Corporation, and the American Chemical Society. He is a Research Career Development Awardee of the National Institutes of Health, and in 1983 and 1984 he was a Visiting Scientist at the Aarhus University Institute of Physiology, Aarhus, Denmark. He has taught biochemistry and physical chemistry at the University of Virginia for 20 years. He is a member of the American Society for Biochemistry and Molecular Biology.

About the Authors

Contents in Brief

For ease in consulting the parent textbook, Biochemistry, *its Table of Contents follows.*

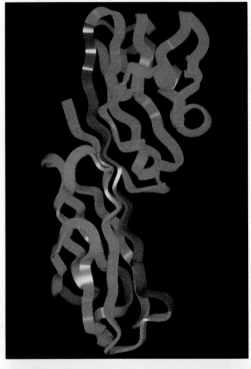

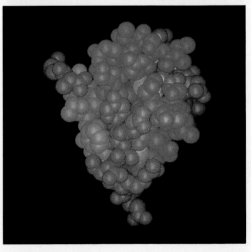

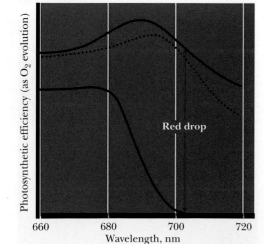

Table of Contents

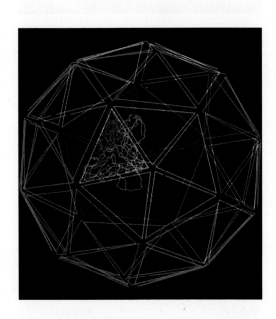

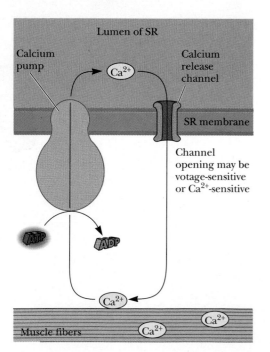

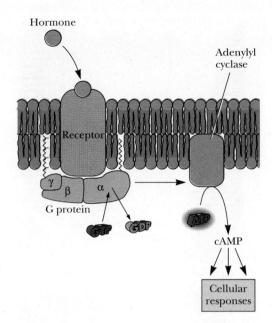

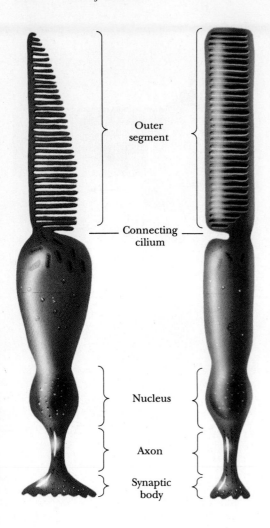

Outer
segment

Connecting
cilium

Nucleus

Axon

Synaptic
body

List of Boxes

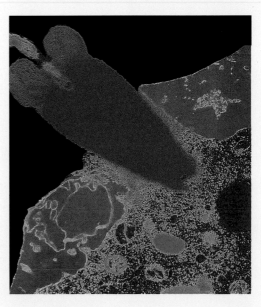

Synopsis of Icon and Color Use in Illustrations

The following symbols and colors are used in this text to help
in illustrating structures, reactions, and biochemical principles:

Elements:

= Oxygen   = Nitrogen   = Phosphorus   = Sulfur   = Carbon   = Chlorine

**Small molecules and groups, which are common reactants
or products in many biochemical reactions, which are
symbolized by the following icons:**

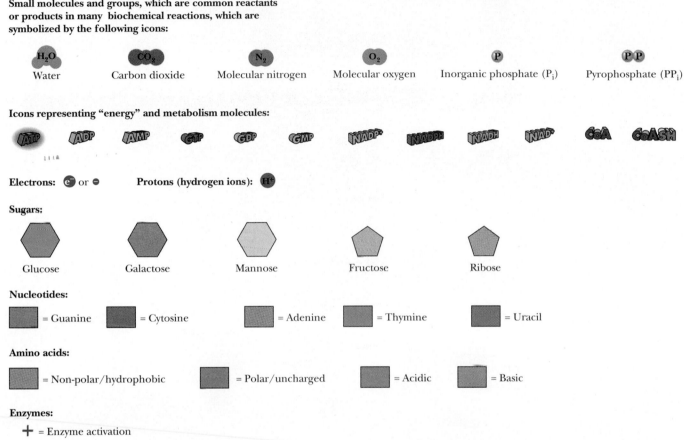

H_2O — Water CO_2 — Carbon dioxide N_2 — Molecular nitrogen O_2 — Molecular oxygen   P — Inorganic phosphate (P_i)   P P — Pyrophosphate (PP_i)

Icons representing "energy" and metabolism molecules:

ATP   ADP   AMP   CTP   GDP   GMP   NADP$^+$ NADPH   NADH   NAD$^+$ CoA   CoASH

Electrons: e$^-$ or ●   **Protons (hydrogen ions):** H$^+$

Sugars:

Glucose   Galactose   Mannose   Fructose   Ribose

Nucleotides:

= Guanine   = Cytosine   = Adenine   = Thymine   = Uracil

Amino acids:

= Non-polar/hydrophobic   = Polar/uncharged   = Acidic   = Basic

Enzymes:

+ = Enzyme activation

⊘ = Enzyme inhibition or inactivation

E = Enzyme   = Enzyme   Enzyme

In reactions, blocks of color over parts of molecular structures are used so that discrete
parts of the reaction can be easily followed from one intermediate to another and
it is easy to see where the reactants originate and how the products are produced.

Some examples:

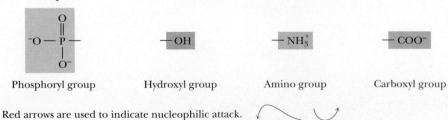

Phosphoryl group   Hydroxyl group   Amino group   Carboxyl group

Red arrows are used to indicate nucleophilic attack.

These colors are internally consistent within reactions and are generally consistent
within the scope of a chapter or treatment of a particular topic.

Part V

Molecular Aspects
of Cell Biology

· ·

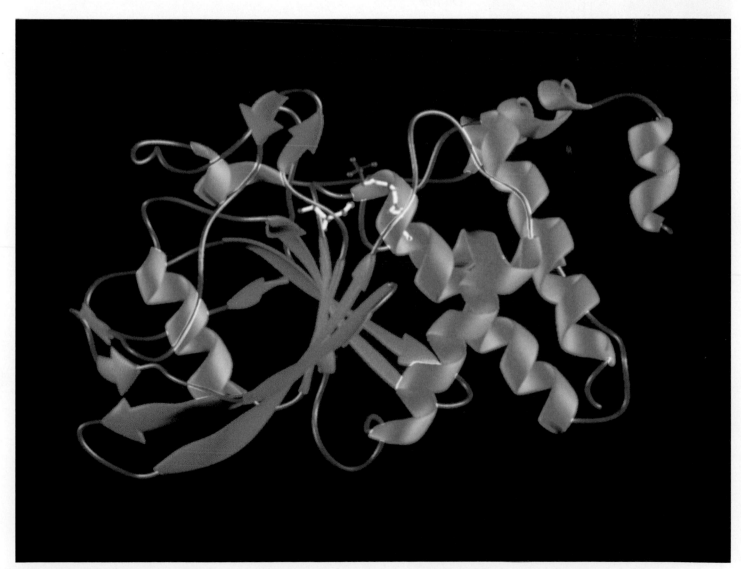

Protein tyrosine phosphatase

Chapter 34

Self-Assembling Macromolecular Complexes

Dallas skyline at night. Apparent similarities between man-made architecture and self-assembling biomolecular structures are superficial, though striking.

A dominant theme in the symphony of life is that complex structures are formed repeatedly and unerringly through the interplay of simple forces and the information coded in the genetic material. Thus, proteins fold into complex tertiary structures dictated by their primary sequences and the relevant hydrogen bonds, as well as electrostatic, hydrophobic, and van der Waals interactions. At the intermolecular level, protein quaternary structures, enzyme-substrate complexes, and DNA double helices likewise can form and maintain their structures without significant intervention by other molecular or cellular entities.

An additional level of structural complexity exists in nature, but it has been considered only briefly to this point. Certain biological molecules possess (in their structures) the ability to form repeating polymeric structures spontaneously. Often these macromolecular complexes perform sophisticated biological functions, but their assembly—in fact, their very existence—depends only on the spontaneous confluence of simple bonding interactions and the information intrinsic in the primary sequences of proteins and nucleic acids. This chapter considers some of these repeating, polymeric structures, the principles that govern their formation, and their complex biological activities. (It should be noted that several other multimeric, self-assembling complexes, including membranes, ribosomes, nucleosomes, snRNPs, and signal recognition particles, have been described in earlier chapters.)

34.1 Basic Principles of Spontaneous Self-Assembly

The structures that are formed in self-assembly processes—microtubules, cytoskeletons, cilia, flagella, and viruses, for example (Figure 34.1)—are incredibly large structures compared with typical globular proteins. If these structures were made from either single polypeptide chains or a large ensemble of different proteins, they would consume enormous amounts of genetic information. Watson and Crick pointed out in the 1950s that typical viruses do not in fact contain enough genetic information to code for viral coats that might consist of a single large molecule or an array of many unique proteins. The construction of a viral coat or a microtubule using repeating structural elements requires much less genetic information. The coding of a viral protein

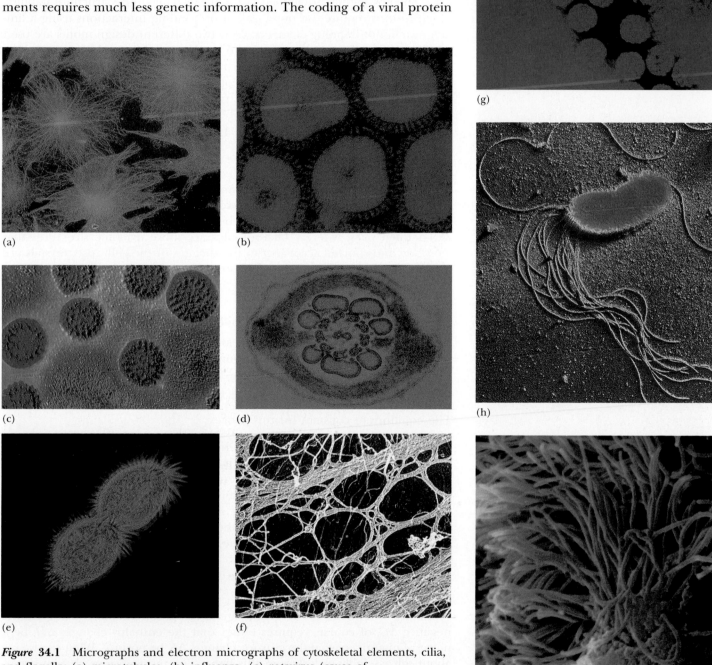

Figure **34.1** Micrographs and electron micrographs of cytoskeletal elements, cilia, and flagella: (a) microtubules, (b) influenza, (c) rotavirus (cause of gastroenteritis), (d) rat sperm tail (cross-section), (e) *Stylonychia*, a ciliated protozoan (undergoing division), (f) cytoskeleton of a eukaryotic cell, (g) polyoma virus, (h) *Pseudomonas fluorescens* (aerobic soil bacterium), showing flagella, (i) nasal cilia.

that may form a coat structure by *self-assembly into a multimeric structure* represents efficient use of the limited genetic information in the viral DNA or RNA.

Large Biological Structures Formed from Repeating Elements Are Highly Symmetric

A second principle important to self-assembly processes concerns the design motifs that can be used to produce large biological structures from repeating elements. At first it may appear that there would be a large variety of ways to efficiently produce such structures. However, this is not at all the case. *Large structures can be built from identical, repeating subunits in only a limited number of ways.* The possibilities are limited by the need to maximize structural stability, which in turn requires an optimization of bonding interactions using a limited number of bonding elements. Only two different design motifs are used consistently in nature—cylinders and polyhedra. Microtubules and many virus coat structures are cylinders formed by helical arrays of one or a few proteins. By contrast, many viruses adopt a nearly spherical (actually polyhedral) structure, with all such known structures having icosahedral symmetry.

Self-Assembling Systems Seek the Lowest Possible Energy State

The driving force for self-assembly processes is the formation of inter-subunit bonds, which create a structure in the lowest possible energy state. In this sense, self-assembly of microtubules and viruses is similar to crystallization. Both these processes are ultimately governed by the laws of statistical thermodynamics, but several differences can be noted. Virus particles grow to finite, well-defined sizes, whereas microtubules are more dynamic structures that lengthen and shorten according to the dictates of the cell. Also, crystals are normally highly uniform, homogeneous structures, but viruses and microtubules consist of at least two chemically and structurally different components.

Equilibrium and Free Energy Changes for Self-Assembly Processes

The energetics of self-assembly processes can be complex, but a few simple points can be made. Assuming the spontaneous formation of a polymer of n subunits—an ***n*-mer,** denoted by X_n—from monomeric species X, we have

$$n\text{X} \longrightarrow \text{X}_n$$

The equilibrium constant for this process is

$$K_n = [\text{X}_n]/[\text{X}]^n$$

The standard free energy change for the formation of an n-mer from a solution of isolated monomer units of X at unit activity is simply

$$\Delta G_n^{\circ\prime} = -RT \ln K_n$$

where R is the gas constant and T is the temperature. Formation of n-mer is more favored as ΔG becomes more negative.

Many Self-Assembly Processes Are Entropy-Driven

The self-assembly of a macromolecule to form a large, complex structure is spontaneous if the free energy change, ΔG, for self-assembly is negative. A negative ΔG, of course, requires either that the enthalpy change, ΔH, be a negative quantity or that the entropy change, ΔS, be positive. What can we say about self-assembly processes in this light? The most obvious change occurring in the assembly of large, symmetric self-assembled structures is that a highly ordered structure has been assembled from a disordered array of subunits. Recalling (Chapter 15) that an increase in order for any process corresponds to a negative ΔS, we might be tempted to argue that the driving force for self-assembly must necessarily arise from a negative ΔH. It is therefore quite surprising to discover that many self-assembly processes are characterized by a positive ΔH. Such is the case for the assembly of tobacco mosaic virus (discussed later), for the polymerization of sickle-cell hemoglobin, and for the formation of collagen fibers from soluble collagen molecules. If this is so, the only way that assembly can be a spontaneous, favored process is if ΔS is also positive (so that ΔG might be negative).

How can ΔS be positive for these (and other) self-assembly processes? The decrease in entropy associated with the formation of the highly ordered, self-assembled structure is more than compensated for by an increase in entropy associated with the solvent water. This increase in ΔS occurs because nonpolar amino acid residues that are exposed to water in the monomeric subunits are "buried" within the structure (at the subunit–subunit interface) when polymerization occurs. As discussed in Chapters 2 and 5, these nonpolar residues, when exposed to solvent, force order on nearby water molecules. Polymerization removes these groups from contact with water, allowing a net disordering of the solvent molecules. This is the very same phenomenon that drives the folding of globular proteins and quaternary associations in oligomeric proteins (Chapter 5).

Substantial experimental evidence supports this model of self-assembly. Stevens and Lauffer showed in 1965 that water molecules are released when tobacco mosaic virus polymerizes. Similar results have since been observed in many other self-assembling systems.

34.2 Tubulin, Microtubules, and Related Structures

One of the simplest self-assembling structures found in biological systems is the *microtubule,* one of the fundamental components of the eukaryotic *cytoskeleton* and the primary structural element of cilia and flagella. **Microtubules** are hollow, cylindrical structures, approximately 30 nm in diameter, formed from **tubulin,** a dimeric protein composed of two similar 55-kD subunits known as *α-tubulin* and *β-tubulin*. Tubulin dimers polymerize as shown in Figure 34.2 to form microtubules, which are essentially helical structures, with 13 tubulin

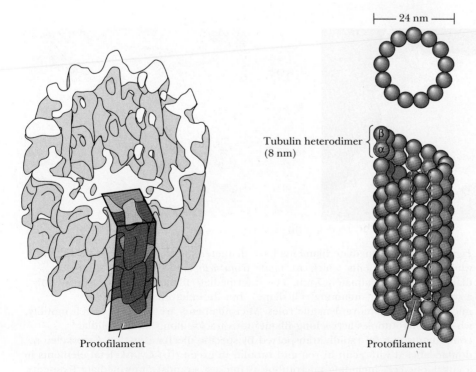

24 nm

Tubulin heterodimer (8 nm)

β
α

Protofilament

Protofilament

Figure **34.2** Microtubules may be viewed as consisting of 13 parallel, staggered protofilaments of alternating α-tubulin and β-tubulin subunits. The sequences of the α and β subunits of tubulin are homologous, and the αβ tubulin dimers are quite stable if Ca^{2+} is present. The dimer is dissociated only by strong denaturing agents.

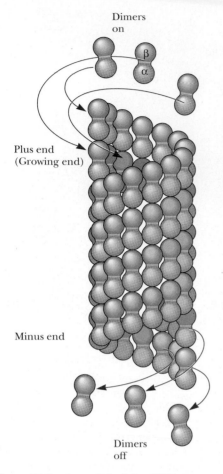

Dimers
on

β

α

Plus end
(Growing end)

Minus end

Dimers
off

***Figure* 34.3** A model of the GTP-dependent treadmilling process. Both α- and β-tubulin possess two different binding sites for GTP. The polymerization of tubulin to form microtubules is driven by GTP hydrolysis, in a process that is only beginning to be understood in detail.

monomer "residues" per turn. Microtubules grown *in vitro* are dynamic structures that are constantly being assembled and disassembled. Since all tubulin monomers in a microtubule are oriented similarly, microtubules are polar structures. The end of the microtubule at which growth occurs is the **plus end,** and the other is the **minus end.** Microtubules *in vitro* carry out a GTP-dependent process called **treadmilling,** in which tubulin monomers are added to the plus end at about the same rate at which monomers are removed from the minus end (Figure 34.3).

Microtubules Are Constituents of the Cytoskeleton

Though composed only of 55-kD tubulin subunits, microtubules can grow sufficiently large to span a eukaryotic cell or to form large structures such as cilia and flagella. Inside cells, networks of microtubules play many functions, including formation of the mitotic spindle that segregates chromosomes during cell division, the movement of organelles and various vesicular structures through the cell, and the variation and maintenance of cell shape. Microtubules are, in fact, a significant part of the **cytoskeleton,** a sort of intracellular scaffold formed of microtubules, *intermediate filaments,* and *microfilaments* (Figure 34.4). In most cells, microtubules are oriented with their minus ends toward the centrosome and their plus ends toward the cell periphery. This consistent orientation is important for mechanisms of intracellular transport.

Microtubules Are the Fundamental Structural Units of Cilia and Flagella

As already noted, microtubules are also the fundamental building blocks of cilia and flagella. **Cilia** are short, cylindrical, hairlike projections on the surfaces of the cells of many animals and lower plants. The beating motion of cilia functions either to move cells from place to place or to facilitate the movement of extracellular fluid over the cell surface. Flagella are much longer structures found singly or a few at a time on certain cells (such as sperm cells). They propel cells through fluids. Cilia and flagella share a common design (Figure 34.5). The **axoneme** is a complex bundle of microtubule

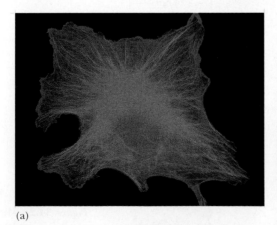

(a)

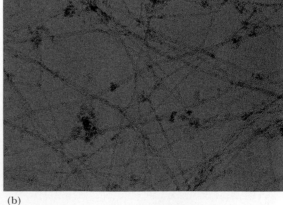

(b)

***Figure* 34.4** Intermediate filaments have diameters of approximately 7 to 12 nm, whereas microfilaments, which are made from *actin* (see Chapter 37), have diameters of approximately 7 nm. The intermediate filaments appear to play only a structural role (maintaining cell shape), but the microfilaments and microtubules play more dynamic roles. Microfilaments are involved in cell motility, whereas microtubules act as long filamentous tracks, along which cellular components may be rapidly transported by specific mechanisms. (a) Cytoskeleton, double-labeled with actin in red and tubulin in green. (b) Cytoskeletal elements in a eukaryotic cell, including microtubules (thickest strands), intermediate filaments, and actin microfilaments (smallest strands).

fibers that includes two central, separated microtubules surrounded by nine pairs of joined microtubules. The axoneme is surrounded by a plasma membrane that is continuous with the plasma membrane of the cell. Removal of the plasma membrane by detergent and subsequent treatment of the exposed axonemes with high concentrations of salt releases the *dynein* molecules (Figure 34.6), which form the *dynein arms*.

The Mechanism of Ciliary Motion

The motion of cilia results from the ATP-driven sliding or walking of dyneins along one microtubule while they remain firmly attached to an adjacent microtubule. The flexible stems of the dyneins remain permanently attached to A-tubules (Figure 34.6). However, the projections on the globular heads form transient attachments to adjacent B-tubules. Binding of ATP to the dynein heavy chain causes dissociation of the projections from the B-tubules. These projections then reattach to the B-tubules at a position closer to the minus end. Repetition of this process causes the sliding of A-tubules relative to B-tubules. The cross-linked structure of the axoneme dictates that this sliding motion will occur in an asymmetric fashion, resulting in a bending motion of the axoneme, as shown in Figure 34.7.

Microtubules Also Mediate Intracellular Motion of Organelles and Vesicles

The unique ability of dyneins to effect **mechano-chemical coupling**—i.e., motion coupled with a chemical reaction—is also vitally important *inside* eukaryotic cells, which, as already noted, contain microtubule networks as part of the cytoskeleton. The mechanisms of intracellular, microtubule-based transport of organelles and vesicles were first elucidated in studies of **axons,** the long

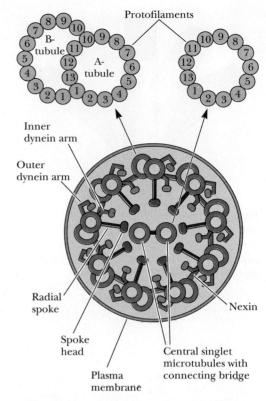

Figure 34.5 The structure of an axoneme. Note the manner in which two microtubules are joined in the nine outer pairs. The smaller-diameter tubule of each pair, which is a true cylinder, is called the **A-tubule** and is joined to the center sheath of the axoneme by a *spoke* structure. Each outer pair of tubules is joined to adjacent pairs by a *nexin* bridge. The **A-tubule** of each outer pair possesses an *outer dynein arm* and an *inner dynein arm*. The larger-diameter tubule is known as the **B-tubule.**

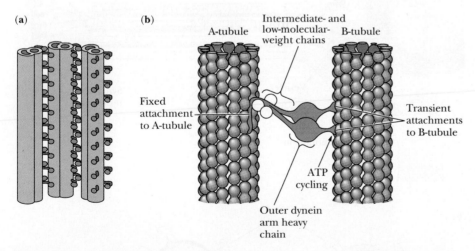

Figure 34.6 (a) Diagram showing dynein interactions between adjacent microtubule pairs. (b) Detailed views of dynein crosslinks between the A-tubule of one microtubule pair and the B-tubule of a neighboring pair. (The B-tubule of the first pair and the A-tubule of the neighboring pair are omitted for clarity.) Isolated axonemal dyneins, which possess ATPase activity, consist of two or three "heavy chains" with molecular masses of 400 to 500 kD, referred to as α and β (and γ when present), as well as several chains with intermediate (40 to 120 kD) and low (15 to 25 kD) molecular masses. Each outer-arm heavy chain consists of a globular domain with a flexible stem on one end and a shorter projection extending at an angle with respect to the flexible stem. In a dynein arm, the flexible stems of several heavy chains are joined in a common base, where the intermediate- and low-molecular-weight proteins are located.

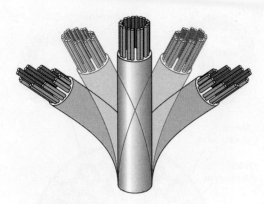

Figure 34.7 A mechanism for ciliary motion. The sliding motion of dyneins along one microtubule while attached to an adjacent microtubule results in a bending motion of the axoneme.

projections of neurons that extend great distances away from the body of the cell. In these cells, it was found that subcellular organelles and vesicles could travel at surprisingly fast rates—as great as 2 to 5 μm/sec—in either direction. Unraveling the molecular mechanism for this rapid transport turned out to be a challenging biochemical problem. The early evidence that these movements occur by association with specialized proteins on the microtubules was met with some resistance, for two reasons. First, the notion that a network of microtubules could mediate transport was novel and, like all novel ideas, difficult to accept. Second, many early attempts to isolate dyneins from neural tissue were unsuccessful, and the dynein-like proteins that were first isolated from cytosolic fractions were thought to represent contaminations from axoneme structures. However, things changed dramatically in 1985 with a report by Michael Sheetz and his co-workers of a new ATP-driven, force-generating protein, different from myosin and dynein, which they called *kinesin*. Then, in 1987, Richard McIntosh and Mary Porter described the isolation of *cytosolic dynein* proteins from *Caenorhabditis elegans,* a nematode worm that never makes motile axonemes at any stage of its life cycle. Kinesins have now been found in many eukaryotic cell types, and similar cytosolic dyneins have been found in fruit flies, amoebae, and slime molds; in vertebrate brain and testes; and in HeLa cells (a unique human tumor cell line).

(a)

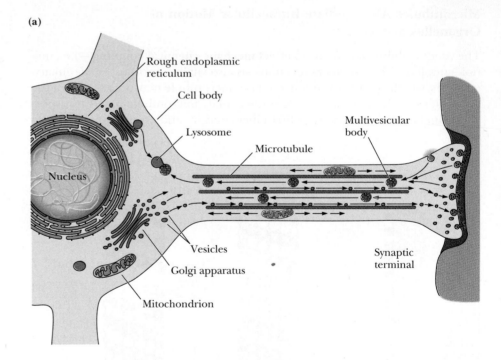

(b)

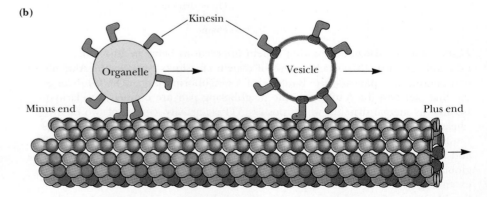

Figure 34.8 Rapid axonal transport along microtubules permits the exchange of material between the synaptic terminal and the body of the nerve cell. Vesicles, multivesicular bodies, and mitochondria are carried through the axon by this mechanism.

Critical Developments in Biochemistry

Microtubule Polymerization Inhibitors as Therapeutic Agents

Microtubules in eukaryotic cells are also important for the maintenance and modulation of cell shape and the disposition of intracellular elements during the growth cycle and mitosis. It may thus come as no surprise that the *inhibition of microtubule polymerization* can block many normal cellular processes. The alkaloid **colchicine** (see figure), a constituent of the swollen, underground stems of the autumn crocus (*Colchicum autumnale*) and meadow saffron inhibits the polymerization of tubulin into microtubules. This effect blocks the mitotic cycle of plants and animals. Colchicine also inhibits cell motility and intracellular transport of vesicles and organelles (which in turn blocks secretory processes of cells). Colchicine has been used for hundreds of years to alleviate some of the acute pain of gout and rheumatism. In gout, white cell lysosomes surround and engulf small crystals of uric acid. The subsequent rupture of the lysosomes and the attendant lysis of the white cells initiate an inflammatory response that causes intense pain. The mechanism of pain alleviation by colchicine is not known for certain, but appears to involve inhibition of white cell movement in tissues. Interestingly, colchicine's ability to inhibit mitosis has given it an important role in the commercial development of new varieties of agricultural and ornamental plants. When mitosis is blocked by colchicine, the treated cells may be left with an extra set of chromosomes. Plants with extra sets of chromosomes are typically larger and more vigorous than normal plants. Flowers developed in this way may grow with double the normal number of petals, and fruits may produce much larger amounts of sugar.

Another class of alkaloids, the **vinca alkaloids** from *Vinca rosea*, the Madagascar periwinkle, can also bind to tubulin and inhibit microtubule polymerization. **Vinblastine** and **vincristine** are used as potent agents for cancer chemotherapy, owing to their ability to inhibit the growth of fast-growing tumor cells. For reasons that are not well understood, colchicine is not an effective chemotherapeutic agent, though it appears to act similarly to the vinca alkaloids in inhibiting tubulin polymerization.

Colchicine

Vinblastine: R = CH₃
Vincristine: R = CHO

The structures of colchicine, vinblastine, and vincristine.

Dyneins Move Organelles in a Plus-to-Minus Direction; Kinesins, in a Minus-to-Plus Direction

The cytosolic dyneins bear many similarities to axonemal dynein. The protein isolated from *C. elegans* includes a "heavy chain" with a molecular mass of approximately 400 kD, as well as smaller peptides with molecular mass ranging from 53 kD to 74 kD. The protein possesses a microtubule-activated ATPase activity, and, when anchored to a glass surface *in vitro,* these proteins, in the presence of ATP, can bind microtubules and move them through the solution. In the cell, cytosolic dyneins specifically move organelles and vesicles from the plus end of a microtubule to the minus end. Thus, as shown in Figure 34.8, dyneins move vesicles and organelles from the cell periphery

(a)

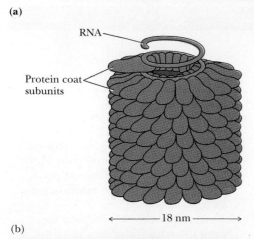

RNA

Protein coat
subunits

←——— 18 nm ———→

(b)

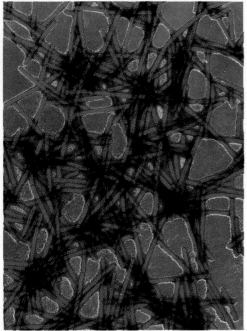

***Figure* 34.9**   (a) A model of a section of a
tobacco mosaic virus (TMV) particle. A
single-stranded RNA molecule forms a right-
handed helix through the core of the
structure, with coat proteins in a helical
array surrounding and protecting the RNA.
(b) An electron micrograph of TMV
particles. The coat protein subunits protect
the RNA. If the protein coat is intact, the
RNA of the virus is insensitive to nucleases,
and the virus can remain infective for many
years. However, if the protein coat is
removed or damaged, the RNA can be
rapidly degraded by chemical and enzymatic
means.

toward the centrosome (or, in an axon, from the synaptic termini toward the
cell body). The **kinesins,** on the other hand, assist the movement of organelles
and vesicles from the minus end to the plus end of microtubules, resulting in
outward movement of organelles and vesicles. Kinesin is similar to cytosolic
dyneins but smaller in size (360 kD), and contains subunits of 110 kD and 65
to 70 kD. Its length is 100 nm. Like dyneins, kinesins possess ATPase activity in
their globular heads, and it is the free energy of ATP hydrolysis that drives the
movement of vesicles along the microtubules.

34.3 Virus Structure and Assembly

All viruses consist of genetic information—in the form of DNA or RNA, but
not both—enclosed and protected by a coat of protein and perhaps lipid.
Viruses are parasitic. They contain the information to reproduce themselves,
but only with the assistance of the biosynthetic machinery of a cell. A given
virus normally can recognize and infect only a particular host (and perhaps
certain closely related species). Virus genetic material ranges from simple to
complex. Satellite tobacco necrosis virus possesses only a single gene in its
genetic material, and the RNA phage called $Q\beta$ contains only four genes. On
the other hand, poxviruses may contain as many as 250 genes.

   The structures of all viruses are designed to balance four fundamental
needs. A virus must (1) be able to protect its genetic material from environmental stresses outside the host cell, (2) be capable of inserting its genetic
material into host cells, (3) replicate its genome, and (4) reproduce its protective structure using the limited amount of genetic information at its dis-
posal. We will examine several viral structures, considering how self-assembling coat and structural proteins enable viruses to meet these four needs.

34.4 Tobacco Mosaic Virus

The first virus to be carefully studied and characterized was **tobacco mosaic
virus (TMV),** a rather simple RNA virus. TMV is a rod-shaped virus 18 nm in
diameter and 300 nm long (Figure 34.9) consisting of a helical, single-
stranded 6390-nucleotide RNA molecule coated with 2130 identical protein
subunits, each interacting with three bases. The mass of each protein subunit
is approximately 17.5 kD, and the mass of the virus particle itself is about 40
million daltons. The overall virus structure is a right-handed helix, with 49
nucleotides and $16\frac{1}{3}$ protein subunits per turn.

Two-layer Disks and Proto-Helices Are Nucleation Points for Formation of Viral Particles

Many important insights into self-assembly processes have been gained from
studies of isolated TMV coat protein. Aaron Klug and co-workers showed in
1971 that the free protein subunits spontaneously form a variety of polymeric
structures, depending on the pH and ionic strength of the solution (Figure
34.10). Small clusters of subunits, circular disks, and stacks of disks predominate above pH 7. However, as the pH is lowered to the range of 6.5 to 7.0, the
protein subunits begin to form **proto-helices**—short helical segments of just
over two turns. Klug's careful ''pH drop'' experiment, in which he diluted a
solution of two-layer disks into a pH 5 acetate buffer (Figure 34.11) and followed the subsequent polymerization of the TMV coat protein, was an important step in the elucidation of the polymerization mechanism. Two-layer disks

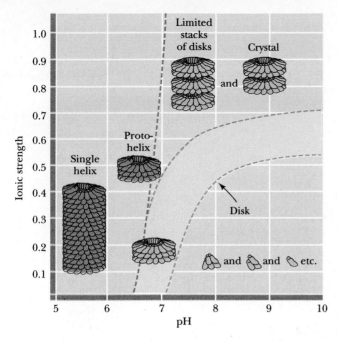

Figure 34.10 Treatment with mild acid dissociates TMV particles into free protein subunits and naked RNA. This figure shows the dependence of TMV coat protein aggregation on ionic strength and pH. At low ionic strength and pH above 7, the predominant structures are small clusters of coat subunits. If the pH is lowered slightly, or if the ionic strength is raised, the protein will spontaneously form into pairs of stacked disks, with 17 subunits per disk. If the ionic strength is increased further, limited stacks of disks, longer stacked disk rods, and crystals of stacked disks are formed.

and protohelices (under some conditions) are nucleation points for the formation of helical rods.

Disks and Proto-Helices Recognize Specific Segments of TMV RNA

Assembly of TMV particles involves specific interactions between proto-helices and TMV RNA. Incubation of dissociated protein subunits with TMV RNA leads to a very slow formation of viral particles. However, incubation of two-layer disks with TMV RNA results in rapid formation of whole virus particles. TMV two-layer disks recognize and associate with a specific segment of TMV RNA (Figure 34.12), forming an intermediate structure from which the entire virus particle may be formed. The crucial segment of RNA contains a base-paired, hairpin structure. The presence of guanine (G) at every third position in a 16-base segment of the hairpin and loop is important, since there is one protein subunit for every three bases.

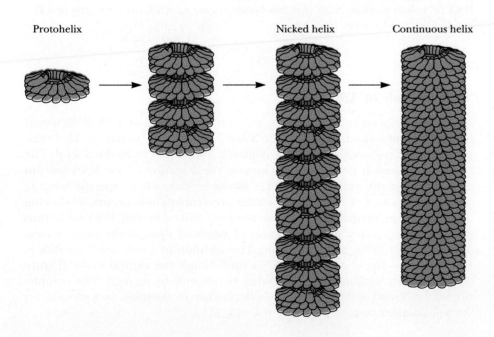

Figure 34.11 Diagram showing the formation (following a "pH drop") of helical arrays of TMV coat protein subunits. The models represent interpretations of the pH drop experiment. Aaron Klug's pH drop experiment showed that disks and proto-helices were nucleation points for the formation of helical TMV particles. The pH drop causes the formation (within 5 seconds) of short, imperfect rods made from two-turn proto-helices separated by misfit regions, so-called "nicked helices." After 15 minutes, the rods are longer and more uniform, and, after 18 hours, the imperfect helices have rearranged themselves into more or less perfect helices.

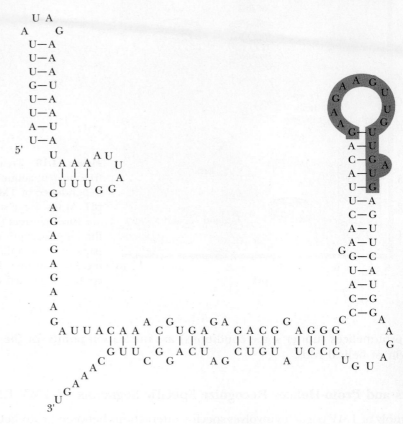

```
                    U  A
                  A      G
                  U — A
                  U — A
                  U — A
                  G — U
                  U — A
                  U — A
                  A — U
                  U — A
            5'    U    A A A U  U
                    A A A    A
                    | | |       A
                    U U U  G  G
                  G
                  A
                  A
                  A
                  A
                  G
                  A
                  A
                  G
                  A U U A C A A   C  U G A   G A C G   A G G G        C
                                  | | |   | | | |   | | | |   | | | |       A
                                  G U U   G A C U   C U G U   U C C C        G
                              A A                                      U G U  A
                            C
                        A A
                      G
                    U
              3'
```

Figure 34.12 The initiation domain on TMV RNA was isolated in an experiment in which small amounts of proto-helices were incubated with native TMV RNA. Under these conditions, proto-helix (RNA) complexes were formed, but since limited amounts of coat protein were added, polymerization to form full virus particles was prevented. "Uncoated" RNA was digested by treatment with ribonuclease. RNA protected through association with proto-helical protein structures was then isolated and sequenced. The sequence (shown here) contained a region that was likely to form a base-paired, hairpin structure with a loop at the end. Interestingly, the hairpin and loop contained a 16-base segment with a guanine (G) at every third position (highlighted in color). This finding attracted immediate attention because each coat subunit binds to three bases on the core RNA of a virus particle. Note that this G-rich region of RNA contains only one C, apparently to discourage base pairing.

The Assembly of TMV

It might have been anticipated that the initiation region of TMV RNA would be located conveniently at either the 3'- or 5'-ends of the molecule. However, the initiation region is found approximately 1000 bases from the 3'-end. The assembly process is now known to involve the insertion of the RNA hairpin and loop into the central cavity of a two-layer disk, where specific binding between the bases of RNA and the coat protein subunits occurs, anchoring the RNA in the central cavity of the two-layer disk. The tiny RNA helix thus formed is propagated by the addition of two-layer disks at the top (i.e., the 5'-end) of the growing virus particle. The addition of a new two-layer disk in this way draws the 5'-end of the RNA up through the central cavity (Figure 34.13), so that additional base–protein bonds may be formed. The completion of the 3'-end of the TMV particle occurs late in assembly, by a process not yet fully understood.

(a)   (b)   (c)

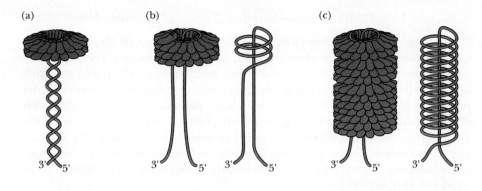

Figure **34.13** TMV RNA threads itself through the coat protein two-layer disk to assemble a virus particle. (a) Insertion of RNA hairpin loop into central cavity of two-layer disk. (b) Intercalation of RNA between the protein subunits converts the disk into a short helix. (c) Successive addition of two-layer disks to the top of the growing virus rod draws the 5′-end of the RNA through the virus. Electron micrographs of growing viral rods show *two* RNA tails extending from one end of each rod. Nuclease digestion experiments demonstrate that, during the assembly process, the 5′-end of the RNA becomes shorter while the 3′-end remains a constant length. These digestion studies also show that 5′-ends of the RNA are protected in incremental packages of about 100 nucleotides at a time. This number corresponds to the number of bases associating with a two-layer disk.

34.5 Spherical and Icosahedral Viruses

The two different self-assembling, macromolecular structures considered thus far (microtubules and TMV particles) both employ helical symmetry motifs. As pointed out earlier, **icosahedral symmetry,** a form of cubic symmetry, is the only other structural symmetry adopted by large self-assembling structures. A regular icosahedron has 20 triangular faces. Each of these triangular faces is threefold symmetric, so that there are a total of 60 (3 × 20) identical and equivalent positions in an icosahedron (Figure 34.14).

Icosahedron

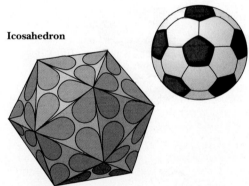

60-subunit icosadeltahedron

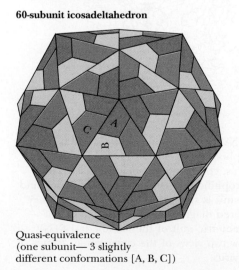

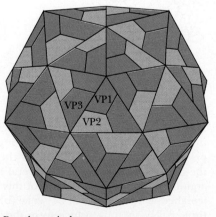

Quasi-equivalence
(one subunit— 3 slightly
different conformations [A, B, C])

Pseudo-equivalence
(3 subunits— VP1, VP2, VP3—
with similar conformations)

Figure **34.14** The structures of an icosahedron and a 60-subunit icosadeltahedron, compared with the icosahedral symmetry of a soccer ball. The case of quasi-equivalent symmetry is shown for a particle in which A, B, and C represent three different conformations of a single polypeptide occupying 180 sites on a virus capsid. Structures with pseudo-equivalent symmetry contain three different subunits with similar conformations.

A Simple Icosahedral Virus—Satellite Tobacco Necrosis Virus

One of the smallest known virus particles, with a radius of about 10 nm, is **satellite tobacco necrosis virus (STNV).** This virus does not even code for proteins needed for its own reproduction in the plant cell. It can multiply only in cells that are simultaneously infected with tobacco necrosis virus. Its protein coat consists of 60 copies of a single protein subunit arranged with classic icosahedral symmetry (Figure 34.14). *This means that each subunit in this virus particle interacts with nearby subunits in exactly the same way. The assembly process thus involves only a single kind of subunit–subunit interaction.* The structure of the subunit (Figure 34.15) consists mainly of two domains of twisted, four-stranded antiparallel β-sheet.

(a)

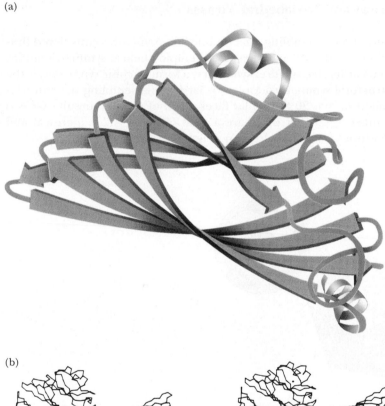

(b)

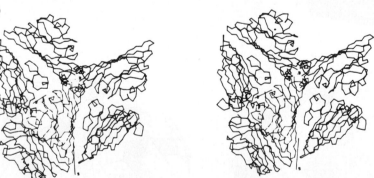

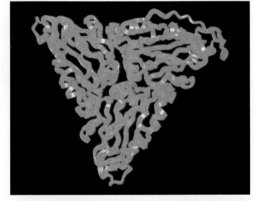

Southern bean mosaic virus (front view)

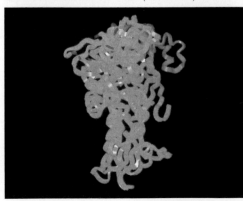

Southern bean mosaic virus (side view)

***Figure* 34.15**   (a) The structure of the STNV protein subunit. The space between the two β-sheets is filled with bulky hydrophobic side chains. The surface of the protein that faces the RNA core of the virus, consisting mainly of β-strands and the N-terminal helix, contains several hydrophilic side chains, including lysine and arginine. The tertiary structure of this subunit is similar to those of several other icosahedral viruses, as will be seen. (b) Stereo diagram of the arrangement of three STNV subunits, which form the asymmetric unit of the icosahedral STNV capsid. The molecular graphics at left show two views of the coat protein trimer of a similar virus, the southern bean mosaic virus.

Tomato Bushy Stunt Virus—A Quasi-Equivalent Icosahedral Structure

The maximum possible number of identical, asymmetric units that can be arranged in a polyhedral structure in which each of the units is identically packed is, in fact, 60. Thus, the icosahedron is the largest truly symmetric structure that can be formed from asymmetric units. Imagine, then, the surprise of the biochemists who discovered that certain "icosahedral" viruses were composed of 180 coat protein subunits. Don Caspar and Aaron Klug pointed out in 1962 that such structures could exist in *approximate* icosahedral symmetry if the subunits could adopt 180/60, or 3, different conformations.

Tomato bushy stunt virus is an example of such a structure. Its coat consists of 180 subunits having identical primary structures that can adopt three distinct and different tertiary structures. A virus structure composed of coat proteins with identical amino acid sequences but different tertiary structures is said to possess **quasi-equivalent symmetry.**

The packing of the protein subunits of tomato bushy stunt virus is illustrated schematically in Figure 34.16. The polyhedron consists of a regular arrangement of subunits with three different tertiary structures, denoted *A*, *B*, and *C*. In turn, each subunit may be viewed (Figure 34.17) as consisting of three domains: **R**, an N-terminal domain involved in interaction with the RNA of the virus core; **S**, the central portion of the peptide that constitutes the *shell;* and **P**, the C-terminal domain that comprises the *projections* observed on

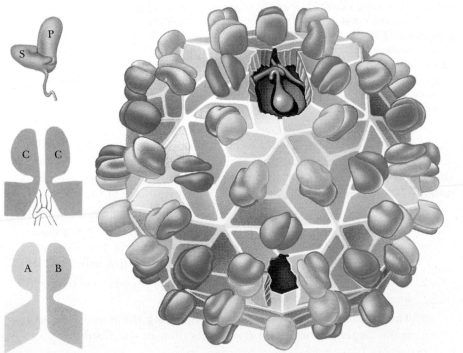
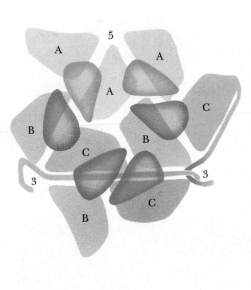

Figure **34.16** The packing of the protein subunits in the TBSV capsid, along with a schematic drawing on the left showing the arrangement of the P and S domains, the interdomain hinge, and the N-terminal arm. On the right is a diagram showing the subunit packing and contacts in the TBSV asymmetric unit (the central triangle). The domains labeled A, B, and C are the S domains of the *A*, *B*, and *C* subunits. The P subunits project outward from the plane of the drawing. For each subunit in the asymmetric unit, the color of the P domain is a lighter shade of the color of its corresponding S domain.

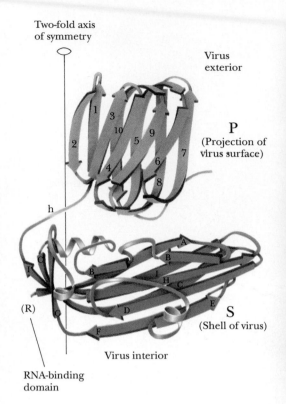

Two-fold axis of symmetry

Virus exterior

P (Projection of virus surface)

h

(R)

S (Shell of virus)

Virus interior

RNA-binding domain

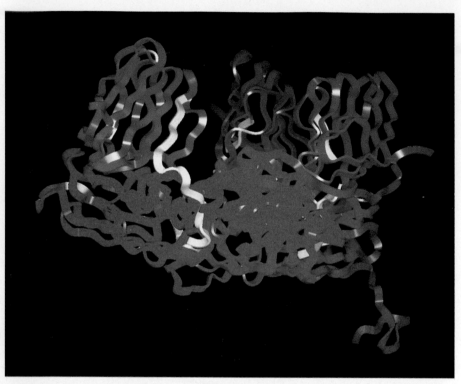

***Figure* 34.17**   The TBSV coat protein subunit consists of 387 amino acid residues and has a molecular mass of approximately 43 kD. The coat is a compact polyhedral shell 3 nm thick, with 90 projections extending along twofold symmetry axes. As shown on the left, the P domain (blue) is composed of two entirely antiparallel β-sheets, one with four strands and one with six. Three β-strands are shared between the two sheets and undergo a tight fold to form a sort of β-sheet sandwich. The S domain (brown) is also largely composed of β-sheet, and is shaped very roughly like a triangular prism. The molecular graphic shows a trimeric array of TBSV coat protein subunits, with P domains in red or purple, S domains in blue, and the hinge domains (h) in white.

the virus surface. An extended arm of 35 residues connects the R and S domains; a short hinge peptide connects S and P. X-ray diffraction studies have determined some of the structural similarities and differences among these subunits that account for the quasi-equivalent symmetry of the TBSV capsid:

1. The P and S domains in each subunit of the viral capsid are folded identically. However, the 35-residue arm is folded in a regular, well-defined way only in the *C* subunits. The arms are disordered and show no well-defined structure in the *A* and *B* subunits.

2. The P domains (projections) of *C* subunits project at a different angle from those of the *A* and *B* subunits.

3. The dihedral angle of the hinge connecting the S and P domains is 30° larger in the *A* and *B* subunits than in the *C* subunits.

A Model for Icosahedral Virus Assembly

A putative model for assembly of icosahedral viruses has come from studies of **turnip crinkle virus (TCV),** another small plant RNA virus that is similar in many ways to TBSV, with a capsid consisting of 180 subunits arranged in the same *ABC* motif. Under conditions of high pH and high ionic strength, TCV particles dissociate into stable fragments. Under these conditions, aggregates of the coat protein and a ribonucleoprotein complex are formed. This *rp-*

complex contains the viral RNA, six coat-protein subunits, and a covalently linked coat-protein dimer. This rp-complex is highly stable and can also be formed from free TCV RNA and coat-protein subunits. The rp-complex does not form with RNA from other sources. Electron microscopy studies show that TCV assembly proceeds by continuous growth of a shell from an initiator that appears to involve the rp-complex, as shown in Figure 34.18.

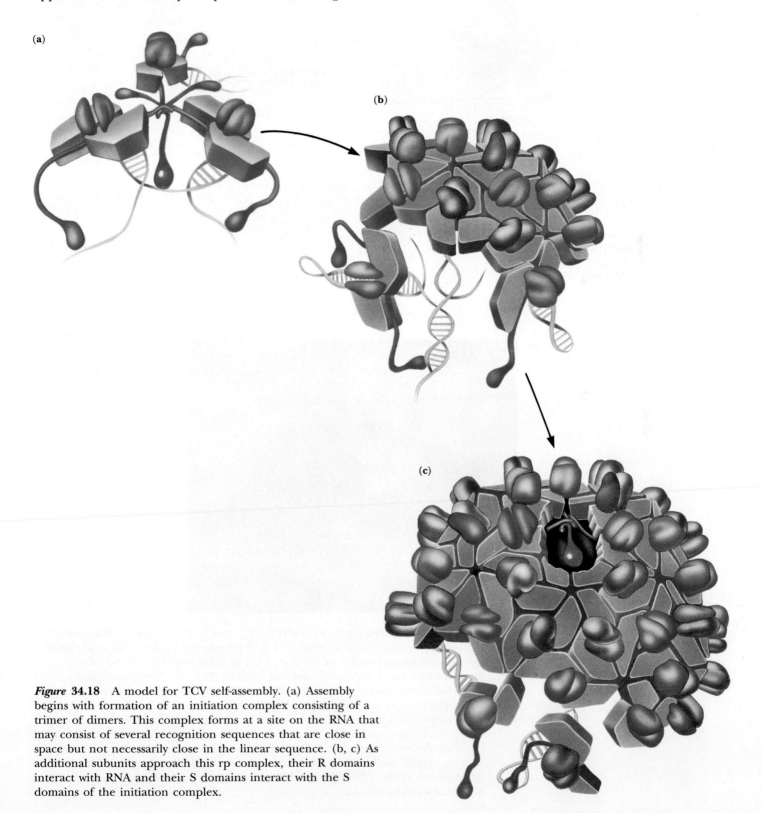

(a)

(b)

(c)

Figure **34.18** A model for TCV self-assembly. (a) Assembly begins with formation of an initiation complex consisting of a trimer of dimers. This complex forms at a site on the RNA that may consist of several recognition sequences that are close in space but not necessarily close in the linear sequence. (b, c) As additional subunits approach this rp complex, their R domains interact with RNA and their S domains interact with the S domains of the initiation complex.

Figure 34.19 (a) The structures of VP1, VP2, and VP3 subunits of rhinovirus are surprisingly similar to one another and to the structures of the other viral capsid proteins considered earlier in the text. Each of these proteins consists of a common core motif with additional elaborations. The cores are topologically identical, each consisting of an eight-stranded antiparallel β-barrel with a pair of short flanking α-helices. The barrel is formed from two twisted β-sheets, and the short loops that join these two sheets give it the appearance of a triangular wedge or prism. At the ends of the β-strands are various extensions and insertions, which decorate the virus exterior, forming hypervariable loops that are also the binding sites for certain antibodies to the picornaviruses. (b) Molecular graphic showing arrangement of the VP1, VP2, and VP3 subunits on the icosahedral surface of human rhinovirus HRV14.

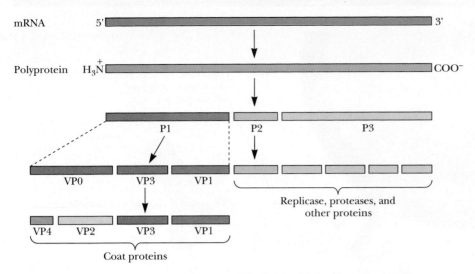

Figure 34.20 Synthesis of rhinovirus proteins by multiple cleavages of a large polyprotein precursor.

Rhinovirus—An Icosahedral Virus That Causes the Common Cold

Rhinovirus, the virus that causes the common cold, is a **picornavirus,** a small, single-stranded RNA-containing animal virus. Picornaviruses derive their name from *pico,* for small, and *rna,* for their genetic material. Picornaviruses have a molecular mass of approximately 8500 kD, of which about 30% is RNA. The virus particles have an external diameter of approximately 30 nm. The coat protein forms an icosahedral shell with 60 protomeric units, each composed of a single copy of four different subunits, usually referred to as VP1, VP2, VP3, and VP4, with molecular masses of approximately 34 kD, 30 kD, 25 kD, and 7 kD, respectively (Figure 34.19). These four proteins are the cleavage products of a single large "polyprotein" encoded by a single strand of messenger RNA. The polyprotein is processed by stepwise proteolysis (Figure 34.20) into several smaller peptide sequences, one of which, called P1, is cleaved further by virally coded proteases to yield the four component protein subunits. The other components of the starting polyprotein become replicases and proteases necessary for viral gene expression and assembly. The most external and antigenic capsid protein of picornaviruses is VP1, whereas VP4 is entirely internal and not exposed to the capsid surface.

34.6 Membrane-Coated Viruses

The viruses considered so far in this chapter are composed entirely of virally coded protein and nucleic acid—no host cell components are incorporated into them. On the other hand, there are many viruses whose coats are lipid membranes "stolen" from the host cells. These viruses are typically assembled in complex processes that culminate in the combination of virally coded proteins with a lipid membrane derived from the host. The assembly of these structures occurs simultaneously with secretion of the viral particle from the host cell. One such membrane-coated virus, **influenza,** has plagued mankind for many centuries. In modern times, more than 20 million people worldwide were killed in the influenza outbreak of 1918, and more recently the Asian flu in 1957 and the Hong Kong flu in 1968 affected smaller numbers of people.

Critical Developments in Biochemistry

Michael Rossman's Canyon Hypothesis

Certain viruses, including human rhinovirus, possess the remarkable ability to consistently evade the immune surveillance of humans or other animals while simultaneously retaining the ability to bind to their cellular receptors through many generations of mutations. How is this possible? Part of the answer arises from the dramatic difference between animal viruses and plant viruses. The plant viruses, which of course are not challenged by host immune systems, have either smooth surfaces or significant projections from the capsid surface. On the other hand, animal virus surfaces typically display depressions. Michael Rossman's **canyon hypothesis** suggests that the depressions, or canyons, on the viral surface are the sites of attachment of cellular receptors, which must be long and slender structures in order to insert into the canyons. Thus, amino acid residues in the lining of the canyon would be highly conserved, so as to maintain good receptor-binding characteristics. The more exposed sites on the viral capsid, which *are* accessible to immune attack, would be variable and subject to mutation, so as to elude the immune system (see figure).

Rossman and his co-workers have mapped portions of the surface of human rhinovirus and have found considerable support for the canyon hypothesis. Replacement of amino acids lining the canyon floor by mutation does indeed change the ability of the virus to bind to cell membranes, whereas mutations outside the canyon have little effect. Interesting support for the canyon hypothesis has come from studies of a series of anti-rhinovirus drugs under development by the Sterling–Winthrop Research Institute. These compounds inhibit viral uncoating after host cell penetration and, more important, inhibit attachment of the virus to the membrane. As shown in part (b) of the figure, X-ray diffraction studies with several of these drugs show that they bind to an interior hydrophobic pocket within VP1 of HRV14, on the canyon

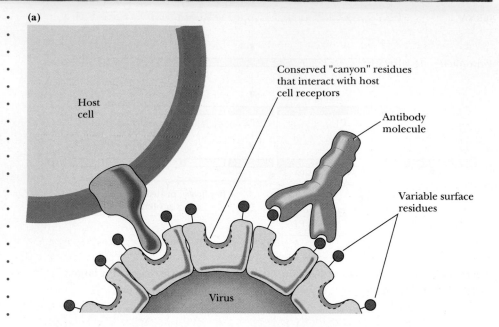

(a)

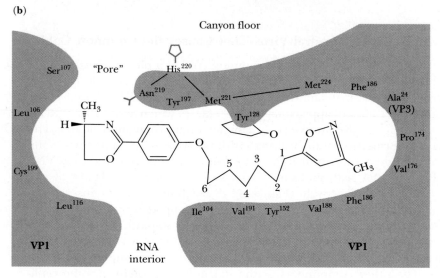

(b)

(a) Canyons on the surface of a rhinovirus particle enable the virus to evade the human immune system. Bulky antibodies cannot penetrate the narrow canyons to bind to conserved residues within the canyon, and the surface residues are varied from one generation of rhinovirus to the next, so that antibodies cannot bind to rhinovirus particles. On the other hand, human cell surface receptors are narrow enough to insert into the canyon and bind tightly to conserved residues on the rhinovirus coat protein. (b) Antivirus compounds bind in the "canyon" of rhinovirus coat protein. In HRV14, the drugs cause conformational changes of up to 0.4 nm in the main-chain positions and 0.75 nm in side-chain positions. The largest changes are in the so-called "FMDV" loop where it crosses the canyon floor, with His^{220} experiencing the biggest change.

floor. The conformational changes induced by drug binding are confined to the canyon itself and do not substantially affect the exterior surface structures of the capsid proteins.

It should be noted that not all picornaviruses use the canyon model for evasion of the host immune system. The structure of foot-and-mouth disease virus shows no obvious surface depressions, and this virus appears to achieve immune system evasion by masking the host receptor-binding site with a flexible peptide loop.

AIDS and the Human Immunodeficiency Virus

Human immunodeficiency virus type 1 (HIV-1), another membrane-coated virus, is the causative agent in **acquired immune deficiency syndrome (AIDS),** a disease transmissible through exchange of bodily fluids. HIV-1 is a membrane-coated virus that binds to and invades **T4 lymphocytes,** a class of white blood cells also known as **helper T cells.** HIV infection progressively destroys whole subsets of the T4 cells, slowly weakening the entire immune system. The development of AIDS makes the patient susceptible to infections and to forms of cancer that are ordinarily suppressed by the immune system. AIDS victims die from these infections and cancers and not (strictly speaking) from the disease itself. So far, AIDS has proven to be 100% fatal.

The Structure of HIV

HIV is a spherical particle, approximately 100 nm in diameter. It consists of a lipid bilayer membrane, derived from the host cell, surrounding an inner particle (Figure 34.21). The membrane envelope contains two glycoproteins, designated **gp120** and **gp41** (*glycoproteins* with masses of 120 kD and 41 kD, respectively), which are together derived from proteolytic cleavage of **gp160,** a larger protein precursor. The gp41 spans the virus membrane, and gp120 extends outward from it, carrying the recognition sites for the host cell. The virus core consists of the RNA along with a few copies of reverse transcriptase and **p7,** a 7-kD nucleocapsid protein. These are surrounded by an elongated conical shell of **p24** proteins (24 kD), which in turn is surrounded by a spherical shell of **p17** proteins (17 kD), which lies adjacent to the lipid membrane envelope.

The Mechanism of HIV Infection

Infection by HIV occurs by attachment of the virus to T4 lymphocytes in the host. Attachment involves binding of gp120 on the virus surface to **CD4,** a

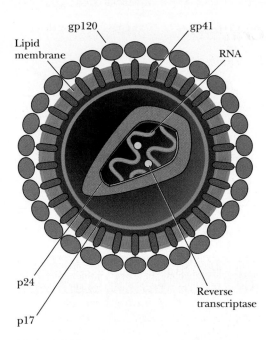

Figure 34.21 The structure of the human immunodeficiency virus (HIV). A lipid membrane derived from the host surrounds the spherical shell composed of p17 proteins, which in turn surrounds a capsule of p24 proteins. Coat protein gp120 is bound noncovalently to the transmembrane coat protein, gp41. The p24 capsule contains RNA and a few copies of reverse transcriptase.

A Deeper Look

Out of Africa? The Origin and Discovery of HIV

In the late 1970s, disturbing reports began to surface of unusual occurrences of a rare form of cancer called Kaposi's sarcoma, a tumor of blood vessel tissue in the skin and internal organs. This disease normally occurs only rarely among older Italian and Jewish men and among natives of Africa. The new reports indicated occurrences of an aggressive form of Kaposi's sarcoma in young, white, middle-class males in the United States and in Europe. The victims also encountered opportunistic infections and a severe depletion of T4 cells, and eventually died from complications resulting from complete breakdown of the immune system. These incidents were the basis for the report by doctors in New York and Los Angeles in 1981 of a new disease syndrome that is now known as **acquired immune deficiency syndrome,** or **AIDS.**

HIV apparently originated in central Africa in the 1950s. HIV antibodies have been detected in blood samples taken from this area at that time. By the late 1970s, HIV had spread to Haiti, and it may have traveled to the Americas and Europe from there. HIV has infected millions of people worldwide, and, unless an effective vaccine can be developed, it may soon threaten millions more.

The rapid spread of AIDS has led to the mobilization of a major research effort directed at an understanding of this disease, its causes, and the strategies for development of possible vaccines and cures. Only three years after AIDS was first described clinically, its cause was shown conclusively to be a human **retrovirus**—a virus that carries RNA as its genetic material. Upon entry into a host, retroviruses use an enzyme known as a **reverse transcriptase** to assemble a double-stranded DNA from the RNA template, degrading the RNA in the process. At the time of its discovery by Luc Montagnier (of the Pasteur Institute in Paris), it was only the third human retrovirus to be identified.

(a) CD4   **(b)** IgG-γ1 Heavy chain

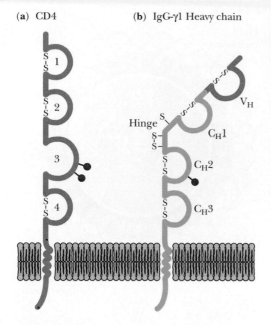

Figure **34.22**   A comparison of the sequences of (a) CD4 and (b) a typical immunoglobulin. CD4 consists of a large extracellular segment (residues 1–371), a single transmembrane span (372–395), and a short, 38-residue, C-terminal cytoplasmic tail (residues 396–433). The extracellular domain of CD4 contains four immunoglobulin V-type domains, designated variously as D1 through D4. Domain D1 has been shown to be necessary and sufficient for gp120 binding.

55-kD cell-surface glycoprotein found mainly on T4 lymphocytes and macrophages. The binding of gp120 to CD4 is apparently followed by fusion of the viral envelope with the cell membrane and entry of the viral genome into the lymphocyte.

The CD4 molecule is a member of the **immunoglobulin gene superfamily** (Figure 34.22), a large class of molecules that generally serve in cellular recognition processes. The normal function of CD4 is to associate with the **class II major histocompatibility complex** on the surfaces of antigen-presenting cells. This complex is involved in the recognition of foreign proteins by the immune system. Binding of CD4 normally facilitates antigen recognition by the T cell receptor, an event that is crucial to the mediation of an efficient cellular immune response.

Blocking viral attachment by inhibiting gp120–CD4 binding is one strategy for the treatment of AIDS. For this reason, the domains of gp120 and CD4 involved in their binding interaction have been the subjects of intense study. In 1990, Wayne Hendrickson and Stephen Harrison's laboratories, working independently, reported the crystal structures of several polypeptide fragments of CD4 (Figure 34.23).

The Mechanism of HIV Assembly

The assembly of virulent HIV particles involves a unique, if inadvertent, cooperation between the host cell and the virus. All the proteins in the HIV particle derive from three protein-coded regions, denoted as *gag*, *pol*, and *env*, in the viral genome. In the host cell, each of these genes codes for a polyprotein (Figure 34.24). The glycoprotein gp160, derived from the *env* gene, is cleaved by a host cell protease, forming gp41 and gp120, which are transported to their destinations within and outside the host plasma membrane, respectively (Figure 34.25). Clustering in the membrane, the gp41–gp120 conjugates act as nucleation sites for the formation of new virus particles.

The viral *gag* and *gag–pol* polyproteins are post-translationally modified by the addition of myristoyl groups on N-terminal glycine residues. These *lipid anchors* (Chapter 9) facilitate attachment of the polyproteins to the plasma

Figure **34.23**   The three-dimensional structure of CD4 residues 1 through 182 consists almost entirely of β-strands arranged in a ''β-sandwich.'' Domain D1 (residues 1–98, blue) consists of nine β-strands, and domain D2 (residues 99–173, yellow) has seven β-strands. The last strand of D1 continues directly into the first strand of D2, resulting in a remarkable 4.9-nm β-strand made up of residues 88 through 103 (red). The interdomain contacts are largely hydrophobic. Mutations that affect the binding of gp120 cluster near the edge of domain 1. The exposed residues that affect gp120 binding decorate one edge of the D1 domain. Several buried residues affect gp120 binding, apparently causing local perturbations in the folded structure. The shape of the D1 domain near these critical residues implies that the complementary shape of the binding domain in gp120 may be a groove rather than a flat surface.

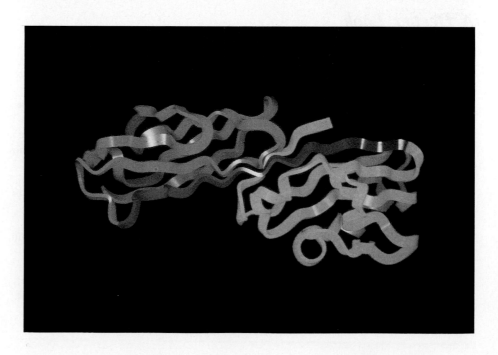

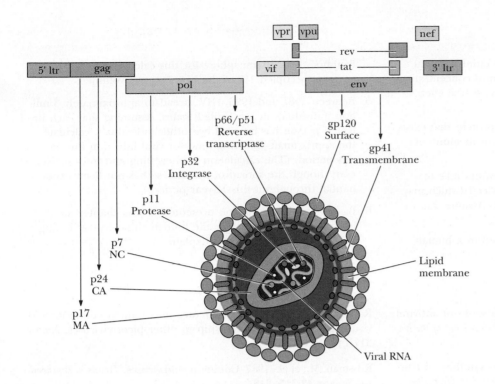

Figure 34.24 The genome of HIV and the polyproteins and proteins it encodes. (NC = nucleocapsid, CA = capsid, and MA = matrix.)

membrane. Once membrane-anchored, these polyproteins associate with viral RNA molecules at the gp41–gp120 nucleation sites, and portions of the plasma membrane pinch off to form immature (noninfectious) virus particles (Figure 34.25).

All subsequent steps in the assembly of HIV occur within these immature virus particles. HIV protease first catalyzes its own release from the *gag–pol* polyprotein and then carries out additional hydrolytic cleavages of the *gag* and *gag–pol* polyproteins to form the complete set of proteins required for the formation of mature virus particles. The capsid protein p17 remains attached to the viral membrane by virtue of its N-terminal myristoyl anchor. The other proteins, including p7 and reverse transcriptase, together with two RNA molecules, assemble within the conical, bullet-shaped viral core formed by p24 (Figure 34.25). Hundreds of mature, infectious HIV particles are produced by a single human T4 cell before it dies from the effects of this process.

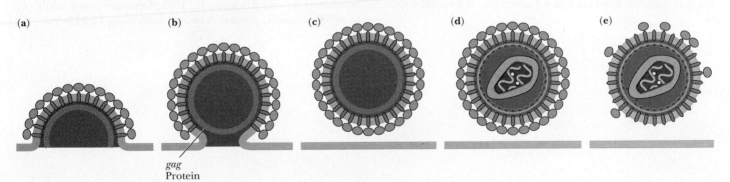

Figure 34.25 A model for the assembly of HIV. (a) The *gag* and *gag–pol* polyproteins are anchored to the host cell membrane by N-terminal myristoyl groups. RNA and nucleocapsid proteins associate in the developing particle. (b) The shell of the core is closed. (c) The HIV protease cleaves the polyproteins in the immature virus particle. (d) The cleaved proteins rearrange to complete the formation of the particle. (e) The mature virus particle has a core–envelope link at the narrow end of the core. HIV particles lose gp120 subunits as they age.

Problems

1. A certain homotetrameric protein exists in equilibrium at 37°C at a concentration of 1 mM with a 1 mM concentration of its monomer subunits. What is the standard free energy of association of this tetrameric protein?

2. Repeat the calculation in problem 1 for a protein that exists as a homooctamer and is in equilibrium with its monomer under the same conditions as in problem 1.

3. Calculate how many tubulin monomers it would take to form a microtubule that would span a liver cell (which may be assumed to be a cube 20 μm on a side). Assume that tubulin monomers are approximately spherical.

4. How many HIV particles would be contained in a human lymphocyte (a 10-μm sphere for this calculation) if virus particles occupied 1% of the lymphocyte's volume?

5. Between 1981 and 1993, HIV spread to approximately 3 million individuals in the United States. Assuming that each infected person has infected two other individuals, calculate the approximate doubling time for viral infection during this period. (The calculation is interesting and informative, even though the spread of the disease has not been exponential throughout this 12-year period.)

6. Based on the information presented in this chapter, would it be appropriate to describe the human rhinovirus as having quasi-equivalent symmetry? Explain.

Further Reading

Badger, J., et al., 1988. Structural analysis of a series of antiviral agents complexed with human rhinovirus 14. *Proceedings of the National Academy of Science U.S.A.* **85**:3304–3308.

Caspar, D., and Klug, A., 1962. Physical principles in the construction of regular viruses. *Cold Spring Harbor Symposia in Quantitative Biology* **27**:1–24.

Fraenkel-Conrat, H., and Kimball, P., 1982. *Virology.* Englewood Cliffs, NJ: Prentice-Hall.

Gelderblom, H., 1991. Assembly and morphology of HIV: Potential effect of structure on viral function. *AIDS* **5**:617–638.

Hogle, J., Chow, M., and Filman, D., 1985. Three-dimensional structure of poliovirus at 2.9 Å resolution. *Nature* **229**:1358–1365.

Holmes, K., 1983. Aaron Klug—Nobel Prize for chemistry. *Trends in Biochemical Sciences* **8**:3–5.

Lauffer, Max A., 1975. *Entropy-Driven Processes in Biology.* London: Chapman and Hall.

Namba, K., and Stubbs, G., 1986. Structure of tobacco mosaic virus at 3.6 Å resolution: Implications for assembly. *Science* **231**:1401–1406.

Rossman, M., et al., 1985. Structure of a human common cold virus and functional relationship to other picornaviruses. *Nature* **317**:145–153.

Rossman, M., et al., 1987. Common cold viruses. *Trends in Biochemical Sciences* **12**:313–318.

Rossman, M., and Johnson, J., 1989. Icosahedral RNA virus structure. *Annual Review of Biochemistry* **58**:533–573.

Stevens, C., and Lauffer, M., 1965. Polymerization–depolymerization of tobacco mosaic virus protein. Part IV, The role of water. *Biochemistry* **4**:31–37.

Volker, E. J., 1993. An attack on the AIDS virus: Inhibition of the HIV-1 protease. *Journal of Chemical Education* **70**:3–9.

Walker, R., and Sheetz, M., 1993. Cytoplasmic microtubule-associated motors. *Annual Review of Biochemistry* **62**:429–451.

Weis, W., et al., 1988. Structure of the influenza virus hemagglutinin complexed with its receptor, sialic acid. *Nature* **333**:426–431.

Wlodawer, A., and Erickson, L., 1993. Structure-based inhibitors of HIV-1 protease. *Annual Review of Biochemistry* **62**:543–585.

Chapter 35

Membrane Transport

"It takes a membrane to make sense out of disorder in biology. You have to be able to catch energy and hold it, storing precisely the needed amount and releasing it in measured shares. A cell does this, and so do the organelles inside. . . . To stay alive, you have to be able to hold out against equilibrium, maintain imbalance, bank against entropy, and you can only transact this business with membranes in our kind of world."

Lewis Thomas, "The World's Biggest Membrane," *The Lives of a Cell* (1974)

"Drawbridge at Arles with a Group of Washerwomen" (1888) by Vincent van Gogh

Transport processes are vitally important to all life forms, since all cells must exchange materials with their environment. Cells must obviously have ways to bring nutrient molecules into the cell and ways to send waste products and toxic substances out. Also, inorganic electrolytes must be able to pass in and out of cells and across organelle membranes. All cells maintain **concentration gradients** of various metabolites across their plasma membranes and also across the membranes of intracellular organelles. By their very nature, cells maintain a very large amount of potential energy in the form of such concentration gradients. Sodium and potassium ion gradients across the plasma membrane mediate the transmission of nerve impulses and the normal functions of the brain, heart, kidneys, and liver, among other organs. Storage and release of calcium from cellular compartments controls muscle contraction, and also the response of many cells to hormonal signals. High acid concentrations in the stomach are required for the digestion of food. Extremely high hydrogen ion gradients are maintained across the plasma membranes of the mucosal cells lining the stomach in order to maintain high acid levels in the stomach yet protect the cells that constitute the stomach walls from the deleterious effects of such acid.

In this chapter, we shall consider the molecules and mechanisms that mediate these transport activities. In nearly every case, the molecule or ion transported is water-soluble, yet moves across the hydrophobic, impermeable lipid membrane at a rate high enough to serve the metabolic and physiologic needs of the cell. This perplexing problem is solved in each case by a specific transport protein. The transported species either diffuses through a channel-forming protein or is carried by a carrier protein. Transport proteins are all classed as **integral membrane proteins** (Chapter 9), ranging in size from small peptides to large, multisubunit protein complexes.

Some transport proteins merely provide a path for the transported species, whereas others couple an enzymatic reaction with the transport event. In all cases, transport behavior depends on the interactions of the transport protein not only with solvent water but with the lipid milieu of the membrane as well. The dynamic and asymmetric nature of the membrane and its components (Chapter 9) plays an important part in the function of these transport systems.

From a thermodynamic and kinetic perspective, there are only three types of membrane transport processes: *passive diffusion*, *facilitated diffusion*, and *active transport*. To be thoroughly appreciated, membrane transport phenomena must be considered in terms of thermodynamics. Some of the important kinetic considerations also will be discussed.

35.1 Passive Diffusion

Passive diffusion is the simplest transport process. In passive diffusion, the transported species moves across the membrane in the thermodynamically favored direction without the help of any specific transport system/molecule. For an uncharged molecule, passive diffusion is an entropic process, in which movement of molecules across the membrane proceeds until the concentration of the substance on both sides of the membrane is the same. For an uncharged molecule, the free energy difference between side 1 and side 2 of a membrane (Figure 35.1) is given by

$$\Delta G = G_2 - G_1 = RT \ln \frac{[C_2]}{[C_1]} \tag{35.1}$$

The difference in concentrations, $[C_2] - [C_1]$, is termed the **concentration gradient**, and ΔG here is the **chemical potential difference**. The **rate of flow** or **flux** (J_C) for the uncharged molecule C across the membrane is given by

$$J_C = -P(C_2 - C_1) \tag{35.2}$$

where J_C is the transport rate per unit area and P is the **permeability coefficient** (usually expressed in cm/sec). P is a constant that describes how readily the molecule leaves the water solvent and crosses the hydrophobic barrier presented by the membrane. P depends on three quantities: (a) the **partition coefficient, K,** which is the ratio of the solubility of the molecule in a hydrophobic solvent (similar to the membrane core) to the solubility in water; (b) the **diffusion coefficient, D,** which describes the rate of diffusion of the molecule in the hydrophobic core of the membrane; and (c) **x, the thickness of the membrane.** P is given by

$$P = \frac{KD}{x} \tag{35.3}$$

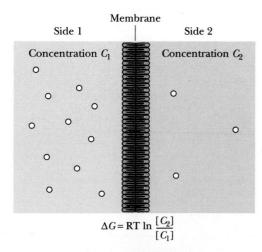

Membrane

Side 1 | Side 2

Concentration C_1 Concentration C_2

$$\Delta G = RT \ln \frac{[C_2]}{[C_1]}$$

***Figure* 35.1** Passive diffusion of an uncharged species across a membrane depends only on the concentrations (C_1 and C_2) on the two sides of the membrane.

Passive Diffusion of a Charged Species

For a charged species, the situation is slightly more complicated. In this case, the movement of a molecule across a membrane depends on its **electrochemical potential.** This is given by

$$\Delta G = G_2 - G_1 = RT \ln \frac{[C_2]}{[C_1]} + Z\mathscr{F}\Delta\psi \qquad (35.4)$$

where Z is the **charge** on the transported species, $\mathscr{F}$ is **Faraday's constant** (the charge on 1 mole of electrons = 96,485 coulombs/mol = 96,485 joules/volt/mol, since 1 volt = 1 joule/coulomb), and $\Delta\psi$ is the electric potential difference (that is, voltage difference) across the membrane. The second term in the expression thus accounts for the movement of a charge across a potential difference. Note that the effect of this second term on ΔG depends on the magnitude and the sign of both Z and $\Delta\psi$. For example, as shown in Figure 35.2, if side 2 has a higher potential than side 1 (so that $\Delta\psi$ is positive), for a negatively charged ion the term $Z\mathscr{F}\Delta\psi$ makes a negative contribution to ΔG.

In other words, the negative charge is spontaneously attracted to the more positive potential—and ΔG is negative. In any case, if the sum of the two terms on the right side of Equation (35.3) is a negative number, transport of the ion in question from side 1 to side 2 would occur spontaneously. The driving force for passive transport is the ΔG term for the transported species itself.

Molecules and ions move in passive ways across biological membranes, although with typically low rates. As noted above, rates of transport depend on the permeability coefficient of the solute. Table 35.1 presents permeability coefficients for several ions and for glucose. The values of P are all quite small. For hydrophilic molecules and ions, this is due primarily to the very low values of K, the partition coefficient. A notable exception is water, which normally exhibits a large value of P relative to other solutes. Although water is a polar substance, it is nevertheless a small, uncharged species that can distribute itself in lipid bilayers and diffuse readily across the bilayer. Also notable in Table 35.1 is the wide range of values for the permeability of protons. The values measured for protons are strongly dependent on the methods and conditions used, and it is now generally agreed that the diffusion of protons is not the result of a single, simple process. In any case, it is clear that proton diffusion is faster than that of other simple cations. Many models have been proposed to describe such diffusion, including models involving the movement of protons along chains of hydrogen-bonded water molecules in the membrane (as in Figure 2.11).

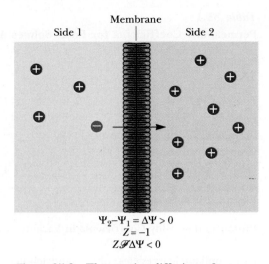

$$\Psi_2 - \Psi_1 = \Delta\Psi > 0$$
$$Z = -1$$
$$Z\mathscr{F}\Delta\Psi < 0$$

Figure 35.2   The passive diffusion of a charged species across a membrane depends upon the concentration and also on the charge of the particle, Z, and the electrical potential difference across the membrane, $\Delta\psi$.

35.2 Facilitated Diffusion

The transport of many substances across simple lipid bilayer membranes via passive diffusion is far too slow to sustain life processes. On the other hand, the transport rates for many ions and small molecules across actual biological membranes is much higher than anticipated from passive diffusion alone. For example, in Table 35.1, the permeability coefficients for chloride ion and glucose in human red blood cells are 10^6 to 10^7 larger than those for the same species in simple bilayer membranes. These differences are due to specific proteins in the red blood cell membrane that **facilitate** transport of these species across the membrane. Similar proteins capable of effecting **facilitated diffusion** of a variety of solutes are present in essentially all natural membranes. Such proteins have two features in common: (a) they facilitate net

Table 35.1

Permeability Coefficients for Polar Solutes Across Bilayers and Biomembranes

Membrane	Permeability Coefficient (cm/sec)				
	Na$^+$	K$^+$	Cl$^-$	H$_2$O	Glucose
Phosphatidylcholine (egg)	$<1.2 \times 10^{-14}$	—	5.5×10^{-11}	4.4×10^{-3}	2.5×10^{-10}
Phosphatidylserine	$<1.6 \times 10^{-13}$	$<9 \times 10^{-13}$	1.5×10^{-11}	—	4×10^{-10}
Phosphatidylserine:cholesterol (1:1)	$<5 \times 10^{-14}$	$<5 \times 10^{-14}$	3.7×10^{-12}	—	1.7×10^{-11}
Human erythrocytes	1×10^{-10}	2.4×10^{-10}	1.4×10^{-4}	5×10^{-3}	2×10^{-5}
Frog erythrocytes	1.4×10^{-7}	1.6×10^{-7}	9.5×10^{-8}	—	—
Dog erythrocytes	—	—	—	5×10^{-3}	—

Proton Permeability Coefficients in Various Membranes	**Permeability Coefficient (cm/sec)**
Large unilamellar vesicles, phosphatidylcholine, phosphatidic acid, 98:2	$P_{net}* = 1.4 \pm 1.6 \times 10^{-4}$
Small unilamellar vesicles, phosphatidylcholine	$P_H = 10^{-8}$–10^{-9}
Large unilamellar vesicles, phosphatidylcholine	$P_H = 3 \times 10^{-9}$
Mitochondria	$P_{net} = 10^{-3}$
Frog muscle sarcolemma	$P_{net} = 10^{-3}$
Sarcoplasmic reticulum	$P_{net} = 10^{-3}$
Erythrocyte membrane	$P_{OH} = 2 \times 10^{-4}$ (pH 9)
	$P_{OH} = 4 \times 10^{-1}$ (pH 4)

Source: Adapted from Jain, M., and Wagner, R. 1980. *Introduction to Biological Membranes*. New York: Wiley.

*P_{net} represents the sum of permeability coefficients for H$^+$ and OH$^-$.

movement of solutes only in the thermodynamically favored direction (that is, $\Delta G < 0$), and (b) they display a measurable affinity and specificity for the transported solute. Consequently, facilitated diffusion rates display **saturation behavior** similar to that observed with substrate binding by enzymes (Chapter 11). Such behavior provides a simple means for distinguishing between passive diffusion and facilitated diffusion experimentally. The dependence of transport rate on solute concentration takes the form of a rectangular hyperbola (Figure 35.3), so that the transport rate approaches a limiting value, V_{max}, at very high solute concentration. Such behavior gives rise to linear plots in Lineweaver-Burk and Hanes-Woolf presentations. Apparent values for K_m, the concentration of solute for which $v = \frac{1}{2}V_{max}$, may be taken from the x-intercepts of either of the latter graphs. Figure 35.3 also shows the graphical behavior exhibited by simple passive diffusion. Since passive diffusion involves no formation of a specific transported complex, the plot of rate versus concentration is linear, not hyperbolic. Moreover, the Lineweaver-Burk plot for passive diffusion passes through the origin, consistent with infinitely high K_m and V_{max}. Likewise, the Hanes-Woolf plot of [S]/v versus [S] has a slope of zero, since [S]/v is a constant for all [S].

Glucose Transport in Erythrocytes Occurs by Facilitated Diffusion

Many transport processes in a variety of cells occur by facilitated diffusion. Table 35.2 lists just a few of these. The **glucose transporter** of erythrocytes illustrates many of the important features of facilitated transport systems. Although glucose transport operates variously by passive diffusion, facilitated

Table 35.2
Facilitated Transport Systems

Permeant	Cell Type	K_m (mM)	V_{max} (mM/min)
D-Glucose	Erythrocyte	4–10	100–500
Chloride	Erythrocyte	25–30	
cAMP	Erythrocyte ghosts	0.0047	0.028
Phosphate	Erythrocyte	80	2.8
D-Glucose	Adipocytes	20	
D-Glucose	Yeast	5	
Sugars and amino acids	Tumor cells	0.5–4	2–6
D-Glucose	Rat liver	30	
D-Glucose	*Neurospora crassa*	8.3	46
Choline	Synaptosomes	0.083	
L-Valine	*Arthrobotrys conoides*	0.15–0.75	

Source: Adapted from Jain, M., and Wagner, R., 1980. *Introduction to Biological Membranes.* New York: Wiley.

diffusion, or active transport mechanisms, depending on the particular cell, the **glucose transport system** of erythrocytes (red blood cells) operates exclusively by facilitated diffusion. The erythrocyte glucose transporter has a molecular mass of approximately 55 kD and is found on SDS polyacrylamide electrophoresis gels (Figure 35.4) as **band 4.5.** Typical erythrocytes contain around 500,000 copies of this protein. The active form of this transport protein in the erythrocyte membrane is a trimer. Hydropathy analysis of the

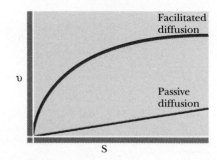
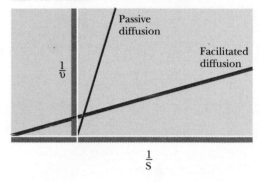
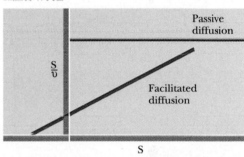

Figure **35.3** Passive diffusion and facilitated diffusion may be distinguished graphically. The plots for facilitated diffusion are similar to plots of enzyme-catalyzed processes (Chapter 11) and they display saturation behavior.

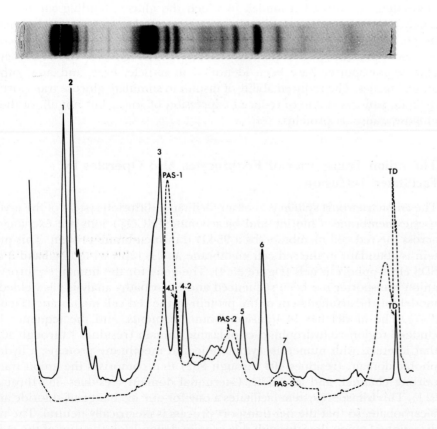

Figure **35.4** SDS-gel electrophoresis of erythrocyte membrane proteins (*top*) and a densitometer tracing of the same gel (*bottom*). The region of the gel between band 4.2 and band 5 is referred to as zone 4.5 or "band 4.5." The bands are numbered from the top of the gel (high molecular weights) to the bottom (low molecular weights). Band 3 is the anion-transporting protein and band 4.5 is the glucose transporter. The dashed line shows the staining of the gel by periodic acid-Schiff's reagent (PAS), which stains carbohydrates. Three "PAS bands" (PAS-1, PAS-2, PAS-3) indicate the positions in the gel of glycoproteins.

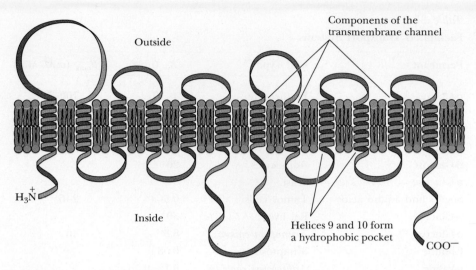

Figure **35.5** A model for the arrangement of the glucose transport protein in the erythrocyte membrane. Hydropathy analysis is consistent with 12 transmembrane helical segments.

amino acid sequence of the erythrocyte glucose transporter has provided a model for the structure of the protein (Figure 35.5). In this model, the protein spans the membrane 12 times, with both the N- and C-termini located on the cytoplasmic side. Transmembrane segments M7, M8, and M11 comprise a hydrophilic transmembrane channel, with segments M9 and M10 forming a relatively hydrophobic pocket adjacent to the glucose-binding site. Cytochalasin B, a fungal metabolite (Figure 35.6), is a competitive inhibitor of glucose transport. The mechanism of glucose transport is not well understood. An **alternating conformation model,** in which the glucose-binding site is alternately exposed to the cytoplasmic and extracellular surfaces of the membrane, has been proposed but remains controversial. Many other glucose transport proteins with sequences that are homologous to the erythrocyte glucose transporter have been identified in muscle, liver, and most other animal tissues. The reduced ability of insulin to stimulate glucose transport in diabetic patients is due to reduced expression of some, but not all, of these glucose transport proteins.

The Anion Transporter of Erythrocytes Also Operates by Facilitated Diffusion

The **anion transport system** is another facilitated diffusion system of the erythrocyte membrane. Chloride and bicarbonate (HCO_3^-) ions are exchanged across the red cell membrane by a 95-kD transmembrane protein. This protein is abundant in the red cell membrane and is represented by **band 3** on SDS electrophoresis gels (Figure 35.4). The gene for the human erythrocyte anion transporter has been sequenced and hydropathy analysis has yielded a model for the arrangement of the protein in the red cell membrane (Figure 35.7). The model has 14 transmembrane segments, and the sequence includes 3 regions: a hydrophilic, cytoplasmic domain (residues 1 through 403) that interacts with numerous cytoplasmic and membrane proteins; a hydrophobic domain (residues 404 through 882) that comprises the anion transporting channel; and an acidic, C-terminal domain (residues 883 through 911). This transport system facilitates a one-for-one exchange of chloride and bicarbonate, so that the net transport process is electrically neutral. The net direction of anion flow through this protein depends on the sum of the chlo-

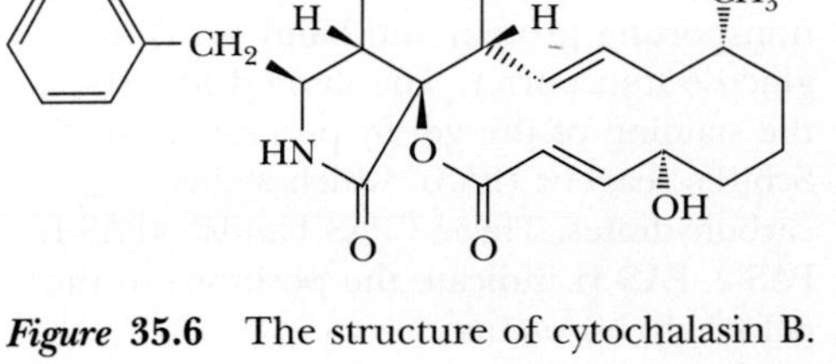

Figure **35.6** The structure of cytochalasin B.

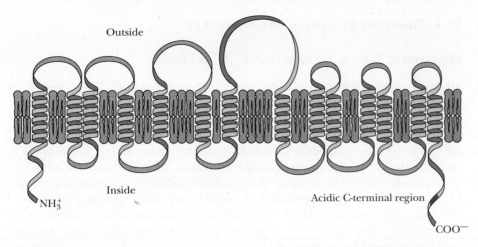

Figure **35.7** Hydropathy plot for the anion transport protein and a model for the arrangement of the protein in the membrane, based on hydropathy analysis.

ride and bicarbonate concentration gradients. Typically, carbon dioxide is collected by red cells in respiring tissues (by means of $Cl^- \rightleftharpoons HCO_3^-$ exchange) and is then carried in the blood to the lungs, where bicarbonate diffuses out of the erythrocytes in exchange for Cl^- ions.

35.3 Active Transport Systems

Passive and facilitated diffusion systems are relatively simple, in the sense that the transported species flow downhill energetically, that is, from high concentration to low concentration. However, other transport processes in biological systems must be *driven* in an energetic sense. In these cases, the transported species move from low concentration to high concentration, and thus the transport requires *energy input*. As such, it is considered an **active transport system.** The most common energy input is **ATP hydrolysis,** with hydrolysis being *tightly coupled* to the transport event. Other energy sources also drive active transport processes, including *light energy* and the *energy stored in ion gradients*. The original ion gradient is said to arise from a **primary active transport** process, and the transport that depends on the ion gradient for its energy input is referred to as a **secondary active transport** process (see discussion of amino acid and sugar transport, Section 35.6). When transport results in a net movement of electric charge across the membrane, it is referred to as an **electrogenic transport** process. If no net movement of charge occurs during transport, the process is electrically neutral.

All Active Transport Systems Are Energy-Coupling Devices

Hydrolysis of ATP is essentially a chemical process, whereas movement of species across a membrane is a mechanical process (that is, movement). An active transport process that depends on ATP hydrolysis thus couples chemical free energy to mechanical (translational) free energy. The bacteriorhodopsin protein in *Halobacterium halobium* couples light energy and mechanical energy. Oxidative phosphorylation (Chapter 20) involves coupling between electron transport, proton translocation, and the capture of chemical energy in the form of ATP synthesis. Similarly, the overall process of photosynthesis (Chapter 22) amounts to a coupling between captured light energy, proton translocation, and chemical energy stored in ATP.

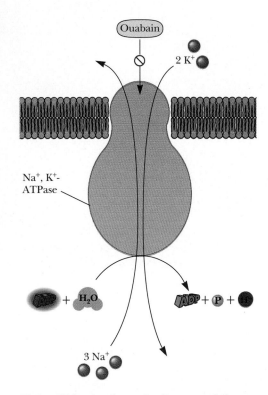

***Figure* 35.8** A schematic diagram of the Na$^+$,K$^+$-ATPase in mammalian plasma membrane. ATP hydrolysis occurs on the cytoplasmic side of the membrane, Na$^+$ ions are transported out of the cell, and K$^+$ ions are transported in. The transport stoichiometry is 3 Na$^+$ out and 2 K$^+$ in per ATP hydrolyzed. The specific inhibitor ouabain and other cardiac glycosides inhibit Na$^+$,K$^+$-ATPase by binding on the extracellular surface of the pump protein.

35.4 Transport Processes Driven by ATP

Monovalent Cation Transport: Na$^+$,K$^+$-ATPase

All animal cells actively extrude Na$^+$ ions and accumulate K$^+$ ions. These two transport processes are driven by **Na$^+$,K$^+$-ATPase,** also known as the **sodium pump,** an integral protein of the plasma membrane. Most animal cells maintain cytosolic concentrations of Na$^+$ and K$^+$ of 10 mM and 100 mM, respectively. The extracellular milieu typically contains about 100 to 140 mM Na$^+$ and 5 to 10 mM K$^+$. Potassium is required within the cell to activate a variety of processes, whereas high intracellular sodium concentrations are inhibitory. The transmembrane gradients of Na$^+$ and K$^+$ and the attendant gradients of Cl$^-$ and other ions provide the means by which neurons communicate (see Chapter 38). They also serve to regulate cellular volume and shape. Animal cells also depend upon these Na$^+$ and K$^+$ gradients to drive transport processes involving amino acids, sugars, nucleotides, and other substances. In fact, maintenance of these Na$^+$ and K$^+$ gradients consumes large amounts of energy in animal cells—20% to 40% of total metabolic energy in many cases and up to 70% in neural tissue.

The Na$^+$- and K$^+$-dependent ATPase comprises two subunits, an α-subunit of 1016 residues (120 kD) and a 35-kD β-subunit. The sodium pump actively pumps three Na$^+$ ions out of the cell and two K$^+$ ions into the cell per ATP hydrolyzed:

$$\text{ATP}^{4-} + \text{H}_2\text{O} + 3\text{Na}^+(\text{inside}) + 2\text{K}^+(\text{outside}) \rightarrow \text{ADP}^{3-} +$$
$$\text{H}_2\text{PO}_4^- + 3\text{Na}^+(\text{outside}) + 2\text{K}^+(\text{inside}) \qquad (35.5)$$

ATP hydrolysis occurs on the cytoplasmic side of the membrane (Figure 35.8), and the net movement of one positive charge outward per cycle makes the sodium pump electrogenic in nature.

Hydropathy analysis of the amino acid sequences of the α- and β-subunits and chemical modification studies have led to a model for the arrangement of the ATPase in the plasma membrane (Figure 35.9). The model describes 10 transmembrane α-helices in the α-subunit, with two large cytoplasmic domains. The larger of these, between transmembrane segments 4 and 5, has been implicated as the ATP-binding domain. The enzyme is covalently phosphorylated at an aspartate residue on the α-subunit in the course of ATP hydrolysis. The covalent E-P intermediate was trapped and identified using tritiated sodium borohydride (Figure 35.10).

A minimal mechanism for Na$^+$,K$^+$-ATPase postulates that the enzyme cycles between two principal conformations, denoted E$_1$ and E$_2$ (Figure 35.11). E$_1$ has a high affinity for Na$^+$ and ATP and is rapidly phosphorylated in the presence of Mg^{2+} to form E$_1$-P, a state which contains *three occluded Na$^+$ ions* (occluded in the sense that they are tightly bound and not easily dissociated from the enzyme in this conformation). A conformation change yields E$_2$-P, a form of the enzyme with relatively low affinity for Na$^+$, but a high affinity for K$^+$. This state presumably releases 3 Na$^+$ ions and binds 2 K$^+$ ions on the outside of the cell. Dephosphorylation leaves E$_2$K$_2$, a form of the enzyme with *two occluded K$^+$ ions.* A conformation change, which appears to be accelerated by the binding of ATP (with a relatively low affinity), releases the bound K$^+$ inside the cell and returns the enzyme to the E$_1$ state. Enzyme forms with occluded cations represent states of the enzyme with cations bound in the transport channel. The alternation between high and low affinities for Na$^+$, K$^+$, and ATP serves to tightly couple the hydrolysis of ATP and ion binding and transport.

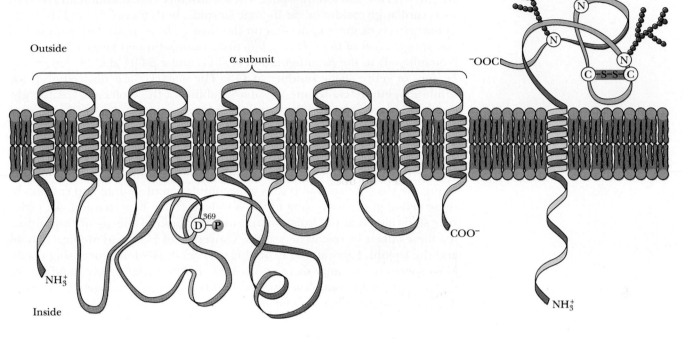

Figure 35.9 A model for the arrangement of Na⁺,K⁺-ATPase in the plasma membrane. The large cytoplasmic domain between transmembrane segments 4 and 5 contains the ATP-binding site and the aspartate residue that is phosphorylated during the catalytic cycle. The β-subunit contains one transmembrane segment and a large extracellular carboxy-terminal segment.

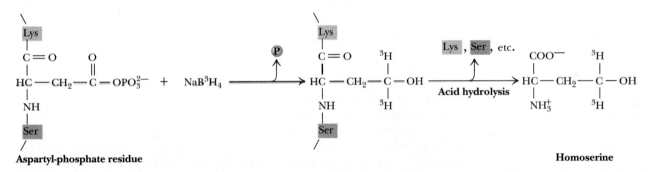

Figure 35.10 The reaction of tritiated sodium borohydride with the aspartyl phosphate at the active site of Na⁺,K⁺-ATPase. Acid hydrolysis of the enzyme following phosphorylation and sodium borohydride treatment yields a tripeptide containing serine, homoserine (derived from the aspartyl-phosphate), and lysine as shown. The site of phosphorylation is Asp³⁶⁹ in the large cytoplasmic domain of the ATPase.

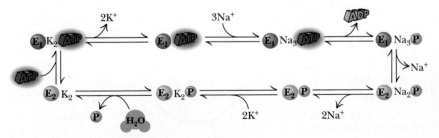

Figure 35.11 A mechanism for Na⁺,K⁺-ATPase. The model assumes two principal conformations, E₁ and E₂. Binding of Na⁺ ions to E₁ is followed by phosphorylation and release of ADP. Na⁺ ions are transported and released and K⁺ ions are bound before dephosphorylation of the enzyme. Transport and release of K⁺ ions completes the cycle.

Na$^+$,K$^+$-ATPase Is Inhibited by Cardiac Glycosides

Plant and animal steroids such as *ouabain* (Figure 35.12) specifically inhibit Na$^+$,K$^+$-ATPase and ion transport. These substances are traditionally referred to as **cardiac glycosides** or **cardiotonic steroids,** both names derived from the potent effects of these molecules on the heart. These molecules all possess a *cis*-configuration of the C-D ring junction, an unsaturated lactone ring (5- or 6-membered) in the β-configuration at C-17, and a β-OH at C-14. There may be one or more sugar residues at C-3. The sugar(s) are not required for inhibition, but do contribute to water solubility of the molecule. Cardiac glycosides bind exclusively to the extracellular surface of Na$^+$,K$^+$-ATPase when it is in the E$_2$-P state, forming a very stable E$_2$-P(cardiac glycoside) complex.

Medical researchers studying high blood pressure have consistently found that people with hypertension have high blood levels of some sort of Na$^+$,K$^+$-ATPase inhibitor. In such patients, inhibition of the sodium pump in the cells lining the blood vessel wall results in accumulation of sodium and calcium in these cells and the narrowing of the vessels to create hypertension. An eight-year study aimed at the isolation and identification of the agent responsible for these effects by researchers at the University of Maryland Medical School and the Upjohn Laboratories in Michigan recently yielded a surprising result. Mass spectrometric analysis of compounds isolated from many hundreds of gallons of blood plasma has revealed that the hypertensive agent is ouabain itself or a closely related molecule!

Calcium Transport: Ca^{2+}-ATPase

Calcium, an ion acting as a cellular signal in virtually all cells (see Chapter 37), plays a special role in muscles. It is the signal that stimulates muscles to contract (Chapter 36). In the resting state, the levels of Ca^{2+} near the muscle fibers are very low (approximately 0.1 μM), and nearly all of the calcium ion in muscles is sequestered inside a complex network of vesicles called the **sarcoplasmic reticulum,** or **SR** (see Figure 36.2). Nerve impulses induce the sarcoplasmic reticulum membrane to quickly release large amounts of Ca^{2+}, with cytoplasmic levels rising to approximately 10 μM. At these levels, Ca^{2+} stimulates contraction. Relaxation of the muscle requires that cytosolic Ca^{2+} levels be reduced to their resting levels. This is accomplished by an ATP-driven Ca^{2+} transport protein known as the **Ca^{2+}-ATPase.** This enzyme is the most abundant protein in the SR membrane, accounting for 70% to 80% of the SR protein. Ca^{2+}-ATPase bears many similarities to the Na$^+$,K$^+$-ATPase. It

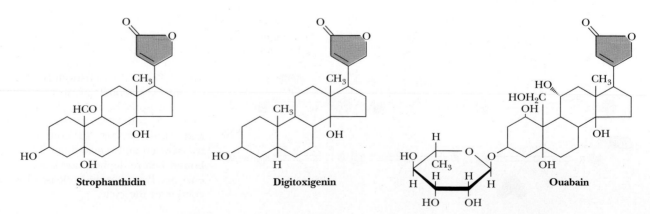

Figure **35.12**   The structures of several cardiac glycosides.

A Deeper Look

Cardiac Glycosides: Potent Drugs from Ancient Times

The cardiac glycosides have a long and colorful history. Many species of plants producing these agents grow in tropical regions and have been used by natives in South America and Africa to prepare poisoned arrows used in fighting and hunting. Zulus in South Africa, for example, have used spears tipped with cardiac glycoside poisons. The sea onion, found commonly in southern Europe and northern Africa, was used by the Romans and the Egyptians as a cardiac stimulant, diuretic, or expectorant. The Chinese have long used a medicine made from the skins of certain toads for similar purposes. Cardiac glycosides are also found in several species of domestic plants, including the foxglove (figure part a), lily of the valley, oleander, and milkweed plant. Monarch butterflies (figure part b) acquire these compounds by feeding on milkweed and then storing the cardiac glycosides in their exoskeletons. Cardiac glycosides deter predation of monarch butterflies by birds, which learn by experience not to feed on monarchs. Viceroy butterflies mimic monarchs in overall appearance. Although viceroys contain no cardiac glycosides and are edible, they are avoided by birds that mistake them for monarchs.

In 1785, the physician and botanist William Withering described the medicinal uses for agents derived from the foxglove plant. In modern times, **digitalis** (a preparation of dried leaves prepared from the foxglove, *Digitalis purpurea*) and other purified cardiotonic steroids have been used to increase the contractile force of heart muscle, to slow the rate of beating, and to restore normal function in hearts undergoing fibrillation (a condition in which heart valves do not open and close rhythmically, but rather remain partially open, fluttering in an irregular and ineffective way). Inhibition of the cardiac sodium pump increases the intracellular Na^+ concentration, leading to stimulation of the Na^+-Ca^{2+} exchanger, which extrudes sodium in exchange for inward movement of calcium. Increased intracellular Ca^{2+} stimulates muscle contraction. Careful use of digitalis drugs has substantial therapeutic benefit for heart patients.

(a)

(b)

(a) Cardiac glycoside inhibitors of Na^+,K^+-ATPase are produced by many plants, including foxglove, lily of the valley, milkweed, and oleander. (b) (*left*) The monarch butterfly, which concentrates cardiac glycosides in its exoskeleton, is shunned by predatory birds. (*right*) Predators also avoid the viceroy, even though it contains no cardiac glycosides, because it is similar in appearance to the monarch.

(a) Phosphorylation domain

	Res. no.																
Na⁺, K⁺-ATPase, α	363	T	S	T	I	C	S	D	K	T	G	T	L	T	Q	N	R	M
H⁺, K⁺-ATPase	379	T	S	V	I	C	S	D	K	T	G	T	L	T	Q	N	R	M
Ca²⁺-ATPase, SR	345	T	S	V	I	C	S	D	K	T	G	T	L	T	T	N	Q	M
H⁺-ATPase, yeast	372	V	E	I	L	C	S	D	K	T	G	T	L	T	K	N	K	L
K⁺-ATPase, *Streptococcus faecalis*	273	L	D	V	I	M	L	D	K	T	G	T	L	T	Q	G	K	F
F₁ ATPase, *E. coli*	280	Q	E	R	I	T	S	T	K	T	G	S	I	T	S	V	Q	A
F₁ ATPase, bovine	293	Q	E	R	I	T	T	T	K	K	G	S	I	T	S	V	Q	A

(b) FITC-reactive region

Na⁺, K⁺-ATPase, α	496	P	Q	H	L	L	V	M	K	G	A	P	E	R	I	L	D	R	C	S	S
H⁺, K⁺-ATPase	510	P	R	H	L	L	V	M	K	G	A	P	E	R	V	L	E	R	C	S	S
Ca²⁺-ATPase, SR	508	V	G	N	K	M	F	V	K	G	A	P	E	G	V	I	D	R	C	N	Y
Ca²⁺-ATPase, plasma membrane					M	Y	S		K	G	A	S	E		I	I	L	R			
H⁺-ATPase, yeast	467	G	E	R	I	V	C	V	K	G	A	P	L	S	A	L	K	T	V	E	E
H⁺-ATPase, *Neurospora*	467	G	E	R	I	T	C	V	K	G	A	P	L	F	V	L	K	T	V	E	E

(c) ATP-binding region

Na⁺, K⁺-ATPase, α	543	L	G	E	R	V	—	L	G	F	C	H	L	F	L	P	D	E	Q	F	P
H⁺, K⁺-ATPase	613	L	K	C	R	T	—	A	G	I	R	V	I	M	V	T	G	D	H	P	I
Ca²⁺-ATPase, SR	611	Q	L	C	R	D	—	A	G	I	R	V	I	M	I	T	G	D	N	K	G
H⁺-ATPase, *Neurospora*	544	C	E	A	K	T	—	L	G	L	S	I	K	M	L	T	G	D	A	V	G
H⁺-ATPase, yeast	545	S	E	A	R	H	—	L	G	L	R	V	K	M	L	T	G	D	A	V	G
F₁ ATPase, bovine	243	E	Y	F	R	D	Q	E	G	G	Q	D	V	L	L	F	I	D	N	I	F	R
F₁ATPase, *E.coli*	267	E	Y	F	R	D	—	R	G	E	D	A	L	I	I	Y	D	D	L	S	K
ATP-ADP exchange protein	277	V	L	—	R	G	N	G	G	A	F	V	L	V	L	Y	D	E	I	K	K
Adenylate kinase	104	E	F	E	R	K	—	I	G	Q	P	T	L	L	L	Y	V	D	A	G	P
Phosphofructokinase	87	E	Q	L	K	K	—	H	G	I	Q	G	L	V	V	I	G	G	D	G	S

(d) Segment binding FSBA and Cl-ATP in Na⁺, K⁺-ATPase

Na⁺, K⁺ATPase, α	701	Q	G	A	I	V	A	V	T	G	D	G	V	N	D	S	P	A	L	K	K
H⁺, K⁺-ATPase	717	L	G	A	I	V	A	V	T	G	D	G	V	N	D	S	P	A	L	K	K
Ca²⁺-ATPase, SR	694	Y	D	E	I	T	A	M	T	G	D	G	V	N	D	A	P	A	L	K	K
H⁺-ATPase, *Neurospora*	625	R	G	Y	L	V	A	M	T	G	D	G	V	N	D	A	P	S	L	K	K
H⁺-ATPase, yeast	625	R	G	Y	L	V	A	M	T	G	D	G	V	N	D	A	P	S	L	K	K
K⁺-ATPase, *Streptococcus faecalis*	467	Q	G	K	K	V	I	M	V	G	D	G	I	N	D	A	P	S	L	A	R

Figure 35.13 Some of the sequence homologies in the nucleotide binding and phosphorylation domains of Na⁺,K⁺-ATPase, Ca²⁺-ATPase, and gastric H⁺,K⁺-ATPase. (FITC is fluorescein isothiocyanate; FSBA is 5-(p-fluorosulfonyl)-benzoyl-adenosine; Cl-ATP is γ-[4-(N-2-chlorethyl-N-methylamino)]-benzoylamid-ATP.)
*(Adapted from Jørgensen, P. L., and Andersen, J. P., 1988. Structural basis for E₁–E₂ conformational transitions in Na⁺,K⁺-pump and Ca-pump proteins. Journal of Membrane Biology **103**:95–120.)*

has an α-subunit of the same approximate size, it forms a covalent E-P intermediate during ATP hydrolysis, and its mechanism of ATP hydrolysis and ion transport is similar in many ways to that of the sodium pump.

The amino acid sequence of the α-subunit is homologous with the sodium pump α-subunit, particularly around the phosphorylation site and the ATP-binding site (Figure 35.13). Ten transmembrane helical segments are predicted from hydropathy analysis, as well as a "stalk" consisting of five helical segments (Figure 35.14). This stalk lies between the membrane surface and the globular cytoplasmic domain containing the nucleotide-binding domain and the site of phosphorylation. The E-P formed by SR Ca²⁺-ATPase is an aspartyl phosphate like that of Na⁺,K⁺-ATPase, in this case residue 351.

Two Ca²⁺ ions are transported into the SR per ATP hydrolyzed by this enzyme, and the mechanism (Figure 35.15) appears to involve two major conformations, E₁ and E₂, just as the Na⁺,K⁺-ATPase mechanism does. Calcium ions are strongly occluded in the E₁-Ca₂-P state, and these occluded ions do not dissociate from the enzyme until the enzyme converts to the E₂-Ca₂-P state, which has a very low affinity for Ca²⁺. In the E₁-Ca₂-P state, the transported Ca²⁺ ions are bound in the transport channel.

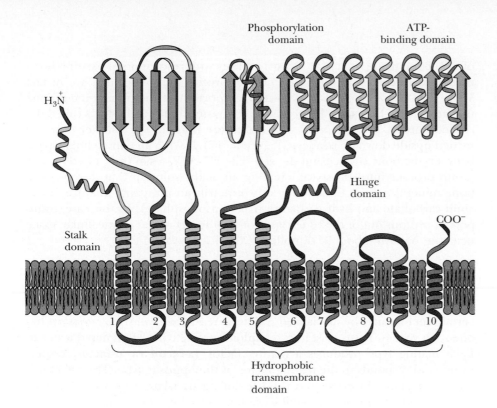

Figure **35.14** The arrangement of Ca^{2+}-ATPase in the sarcoplasmic reticulum membrane. Ten transmembrane segments are postulated on the basis of hydropathy analysis.

The Gastric H⁺,K⁺-ATPase

Production of protons is a fundamental activity of cellular metabolism (Chapter 33), and proton production plays a special role in the stomach. The highly acidic environment of the stomach is essential for the digestion of food in all animals. The pH of the stomach fluid is normally 0.8 to 1. The pH of the parietal cells of the gastric mucosa in mammals is approximately 7.4. This represents a **pH gradient** across the mucosal cell membrane of 6.6, the largest known transmembrane gradient in eukaryotic cells. This enormous gradient must be maintained constantly so that food can be digested in the stomach without damage to the cells and organs adjacent to the stomach. The gradient of H⁺ is maintained by an **H⁺,K⁺-ATPase,** which uses the energy of hydrolysis of ATP to pump H⁺ out of the mucosal cells and into the stomach interior in exchange for K⁺ ions. This transport is electrically neutral, and the K⁺ that is transported into the mucosal cell is subsequently pumped back out of the cell together with Cl⁻ in a second electroneutral process (Figure 35.16). Thus, the net transport effected by these two systems is the movement of HCl into the interior of the stomach. (Only a small amount of K⁺ is needed, since it is recycled.) The H⁺,K⁺-ATPase bears many similarities to the plasma membrane Na⁺,K⁺-ATPase and the SR Ca²⁺-ATPase described above. It has a similar molecular weight, forms an E-P intermediate, and many parts of its peptide sequence are homologous with the Na⁺,K⁺-ATPase and Ca²⁺-ATPase (Figure 35.13).

Bone Remodeling by Osteoclast Proton Pumps

Other proton-ATPases exist in eukaryotic and prokaryotic systems (Chapter 33). **Vacuolar ATPases** are found in vacuoles, lysosomes, endosomes, Golgi, chromaffin granules, and coated vesicles. Various H⁺-transporting ATPases occur in yeast and bacteria as well. H⁺-transporting ATPases found in **osteoclasts** (multinucleate cells that break down bone during normal bone remodeling) provide a source of circulating calcium for soft tissues such as nerves and muscles. About 5% of bone mass in the human body undergoes remodeling at any given time. Once growth is complete, the body balances formation

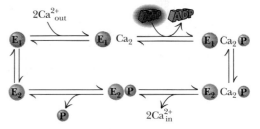

Figure **35.15** A mechanism for Ca²⁺-ATPase from sarcoplasmic reticulum. Note the similarity to the mechanism of Na⁺,K⁺-ATPase (see also Figure 35.11).

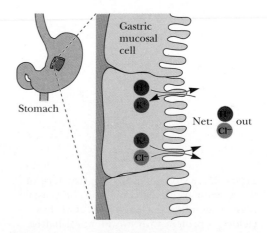

Figure **35.16** The H⁺,K⁺-ATPase of gastric mucosal cells mediates proton transport into the stomach. Potassium ions are recycled by means of an associated K⁺/Cl⁻ cotransport system. The action of these two pumps results in net transport of H⁺ and Cl⁻ into the stomach.

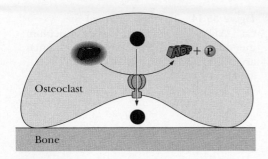

Figure 35.17   Proton pumps cluster on the ruffled border of osteoclast cells and function to pump protons into the space between the cell membrane and the bone surface. High proton concentration in this space dissolves the mineral matrix of the bone.

of new bone tissue by cells called **osteoblasts** with resorption of existing bone matrix by osteoclasts. Osteoclasts possess proton pumps—which are in fact V-type ATPases (see Chapter 33)—on the portion of the plasma membrane that attaches to the bone. This region of the osteoclast membrane is called the ruffled border. The osteoclast attaches to the bone in the manner of a cup turned upside down on a saucer (Figure 35.17), leaving an extracellular space between the bone surface and the cell. The H^+-ATPases in the ruffled border pump protons into this space, creating an acidic solution that dissolves the bone mineral matrix. Bone mineral is primarily an inorganic mixture of calcium carbonate and hydroxyapatite (calcium phosphate). In this case, transport of protons out of the osteoclasts lowers the pH of the extracellular space near the bone to about 4, solubilizing the hydroxyapatite.

ATPases That Transport Peptides and Drugs

Species other than protons and inorganic ions are also transported across certain membranes by specialized ATPases. Yeast (*Saccharomyces cerevisiae*) has one such system. Yeasts exist in two haploid mating types, designated **a** and **α.** Each mating type produces a mating factor (**a-factor** or **α-factor,** respectively) and responds to the mating factor of the opposite type. The α-factor is a peptide that is inserted into the ER during translation on the ribosome, thanks to the presence of a signal sequence. α-Factor is glycosylated in the ER and then secreted from the cell. On the other hand, the a-factor is a 12-amino acid peptide made from a short precursor with no signal sequence or glycosylation site. Export of this peptide from the cell is carried out by a 1290-residue protein, which consists of two identical halves joined together—a tandem duplication (Chapter 33). Each half contains six putative transmembrane segments arranged in pairs, and a conserved hydrophilic cytoplasmic domain

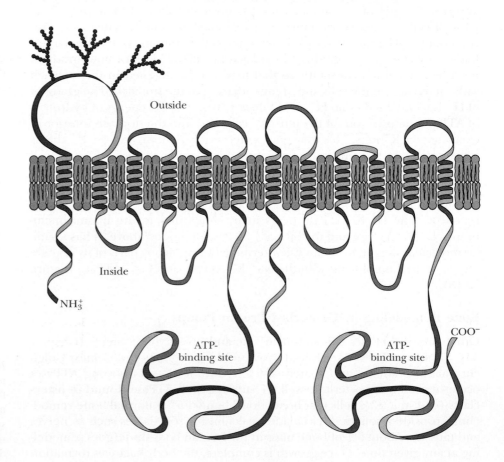

Figure 35.18   A model for the structure of the a-factor transport protein in the yeast plasma membrane. Gene duplication has yielded a protein with two identical halves, each half containing six transmembrane helical segments and an ATP-binding site. Like the yeast a-factor transporter, the multidrug transporter is postulated to have 12 transmembrane helices and 2 ATP-binding sites.

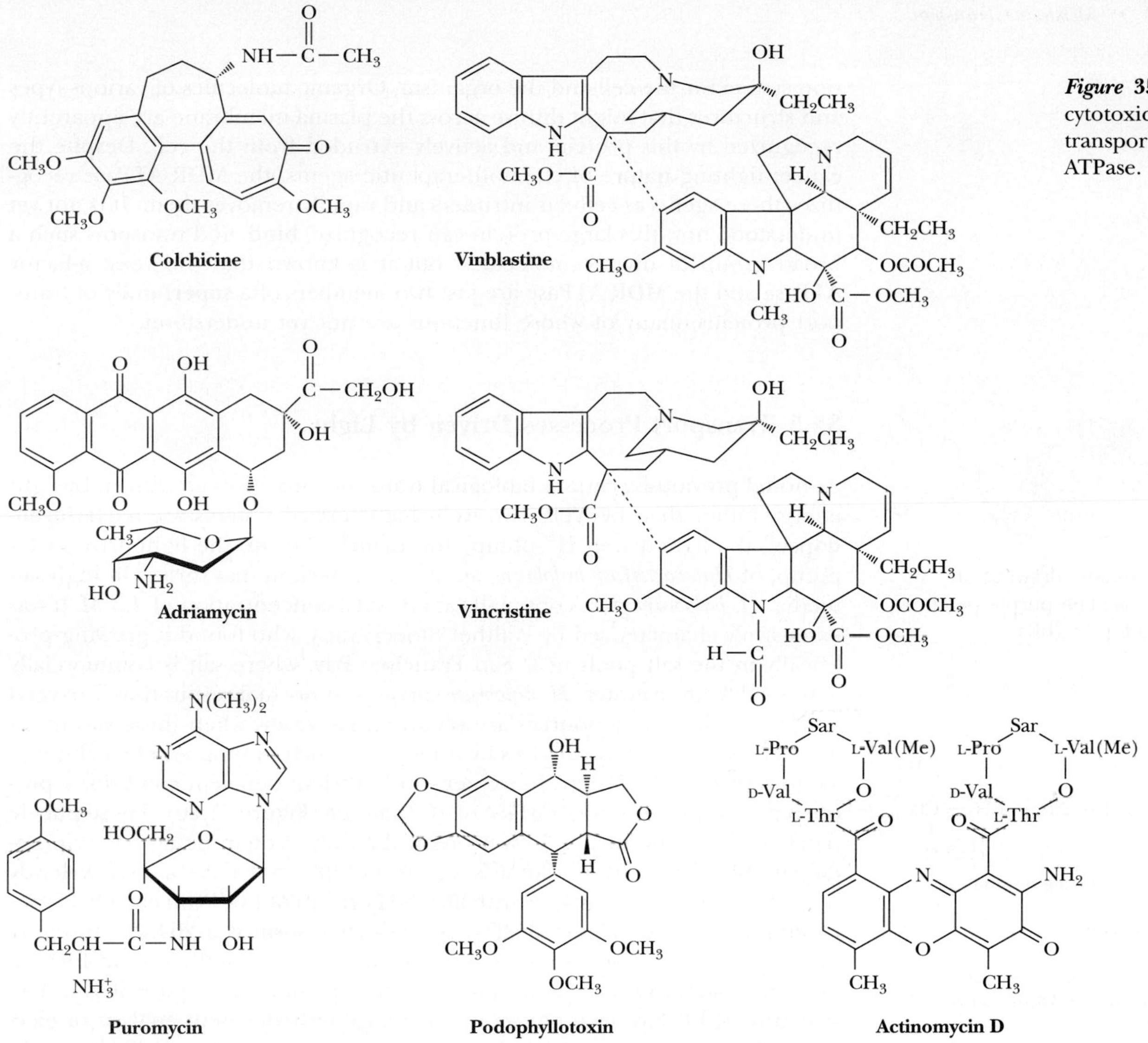

Colchicine Vinblastine Adriamycin Vincristine Puromycin Podophyllotoxin Actinomycin D

Figure **35.19** Some of the cytotoxic drugs that are transported by the MDR ATPase.

containing a consensus sequence for an ATP-binding site (Figure 35.18). This protein uses the energy of ATP hydrolysis to export the 12-residue a-factor from the cell. In yeast cells that produce mutant forms of the a-factor ATPase, a-factor is not excreted and accumulates to high levels inside the cell.

Proteins very similar to the yeast a-factor transporter have been identified in a variety of prokaryotic and eukaryotic cells, and one of these appears to be responsible for the acquisition of **drug resistance** in many human malignancies. Clinical treatment of human cancer often involves chemotherapy, the treatment with one or more drugs that selectively inhibit the growth and proliferation of tumorous tissue. However, the efficacy of a given chemotherapeutic drug often decreases with time, owing to an acquired resistance. Even worse, the acquired resistance to a single drug usually results in a simultaneous resistance to a wide spectrum of drugs with little structural or even functional similarity to the original drug, a phenomenon referred to as **multidrug resistance,** or **MDR.** This perplexing problem has been traced to the induced expression of a 170-kD plasma membrane glycoprotein known as the **P-glycoprotein** or the **MDR ATPase.** Like the yeast a-factor transporter, MDR ATPase is a tandem repeat, each half consisting of a hydrophobic sequence with six transmembrane segments followed by a hydrophilic, cytoplasmic sequence containing a consensus ATP-binding site (Figure 35.18). The protein uses the energy of ATP hydrolysis to actively transport a wide variety of drugs (Figure 35.19) out of the cell. Ironically, it is probably part of a sophisticated protection system for the cell and the organism. Organic molecules of various types and structures that might diffuse across the plasma membrane are apparently recognized by this protein and actively extruded from the cell. Despite the cancer-fighting nature of chemotherapeutic agents, the MDR ATPase recognizes these agents as cellular intruders and rapidly removes them. It is not yet understood how this large protein can recognize, bind, and transport such a broad group of diverse molecules, but it is known that the yeast a-factor ATPase and the MDR ATPase are just two members of a **superfamily** of transport proteins, many of whose functions are not yet understood.

35.5 Transport Processes Driven by Light

As noted previously, certain biological transport processes are driven by light energy rather than by ATP. Two well-characterized systems are **bacteriorhodopsin,** the light-driven H^+-pump, and **halorhodopsin,** the light-driven Cl^--pump, of *Halobacterium halobium,* an archaebacterium that thrives in high-salt media. *H. halobium* grows optimally at an NaCl concentration of 4.3 *M*. It was extensively characterized by Walther Stoeckenius, who found it growing prolifically in the salt pools near San Francisco Bay, where salt is commercially extracted from seawater. *H. halobium* carries out normal respiration if oxygen and metabolic energy sources are plentiful. However, when these substrates are lacking, *H. halobium* survives by using bacteriorhodopsin and halorhodopsin to capture light energy. In oxygen- and nutrient-deficient conditions, **purple patches** appear on the surface of *H. halobium* (Figure 35.20). These purple patches of membrane are 75% protein, the only protein being **bacteriorhodopsin (bR).** The purple color arises from a retinal molecule that is covalently bound in a Schiff base linkage with an ϵ-NH_2 group of Lys^{216} on each bacteriorhodopsin protein (Figure 35.21). Bacteriorhodopsin is a 26-kD transmembrane protein that packs so densely in the membrane that it naturally forms a two-dimensional crystal in the plane of the membrane (Figure 9.37). The structure of bR has been elucidated by image enhancement analysis of electron microscopic data, which reveals seven transmembrane helical protein segments (Figures 9.37 and 9.38). The retinal moiety lies parallel to the membrane plane, about 1 nm below the membrane's outer surface (Figure 9.38).

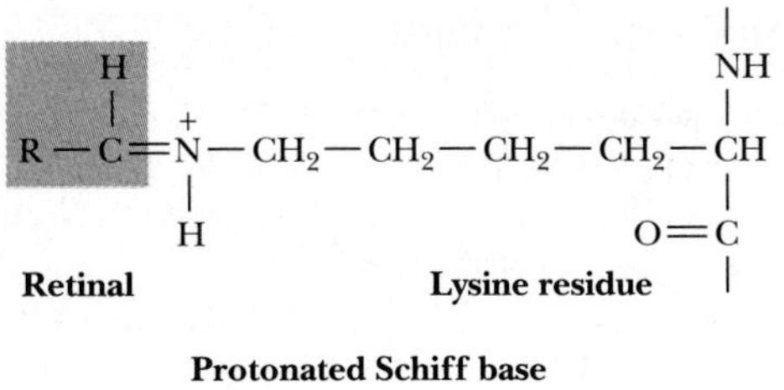

A Model for Light-Driven Proton Transport

The mechanism of the light-driven transport of protons by bacteriorhodopsin is complex, but a partial model has emerged (Figure 35.22). A series of intermediate states, named for the wavelengths (in nm) of their absorption spectra, has been identified. Absorption of a photon of light by the bR_{568} form (in which the Schiff base at Lys^{216} is protonated) converts the retinal from the all-*trans* configuration to the 13-*cis* isomer. Passage through several different intermediate states results in outward transport of $2\ H^+$ ions per photon absorbed, and the return of the bound retinal to the all-*trans* configuration. It appears that the transported protons are in fact protons from the protonated Schiff base. The proton gradient thus established is used by *H. halobium* to drive ATP synthesis and the movement of molecules across the cell membrane. Light-driven ATP synthesis has been demonstrated in an artificial membrane system containing bR (Figure 35.23).

Light-Driven Chloride Transport in *H. halobium*

Anion transport, on the other hand, is handled by a second light-driven ion pump in the *H. halobium* membrane. The inward transport of Cl^- ion is medi-

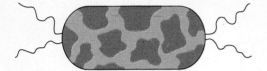

Figure 35.20 A schematic drawing of *Halobacterium halobium.* The purple patches contain bacteriorhodopsin (bR).

Retinal   **Lysine residue**

Protonated Schiff base

Figure 35.21 The Schiff base linkage between the retinal chromophore and Lys^{216}.

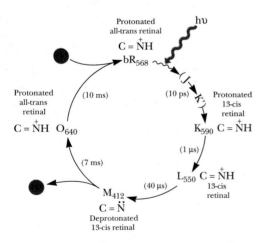

Figure 35.22 The reaction cycle of bacteriorhodopsin. The intermediate states are indicated by letters, with subscripts to indicate the absorption maxima of the states. Also indicated for each state is the configuration of the retinal chromophore (all-*trans* or 13-*cis*) and the protonation state of the Schiff base (C=N: or C=N$^+$H).

ated by *halorhodopsin*, a 27-kD protein whose primary sequence and arrangement in the membrane (Figure 35.24) is very similar to that of bacteriorhodopsin. Although halorhodopsin does not exist naturally as a tightly packed two-dimensional crystal in the membrane, it does have a retinal chromophore, bound covalently at Lys^{242}, the only lysine in the protein. The transmembrane portion of halorhodopsin is 36% homologous with bacteriorhodopsin. The conserved residues are concentrated in the central core formed in both proteins by the seven transmembrane helices (Figure 35.25). Like bacteriorhodopsin, halorhodopsin undergoes a cycle of light-driven conformational changes (Figure 35.26), but no deprotonation of the Schiff base occurs during the halorhodopsin photocycle. Given the striking similarity of structures for these two proteins, it is intriguing to ask why bacteriorhodopsin pumps H^+ but not Cl^- and why halorhodopsin pumps Cl^- but not H^+. The first question may be answered by the work of H. G. Khorana and his co-workers, who replaced Asp^{85} and Asp^{96} in bacteriorhodopsin with asparagine and found that either substitution caused a drastic reduction in H^+ transport. Dieter Oesterhelt and co-workers have shown that Asp^{85} and Asp^{96} are important in the deprotonation and reprotonation, respectively, of the Schiff base in bacteriorhodopsin. The absence of these two crucial residues in halorhodopsin may explain why the latter protein can't reversibly deprotonate the Schiff base and why halorhodopsin doesn't pump protons.

35.6 Transport Processes Driven by Ion Gradients

Amino Acid and Sugar Transport

The gradients of H^+, Na^+, and other cations and anions established by ATPases and other energy sources can be used for **secondary active transport** of various substrates. The best-understood systems use Na^+ or H^+ gradients to

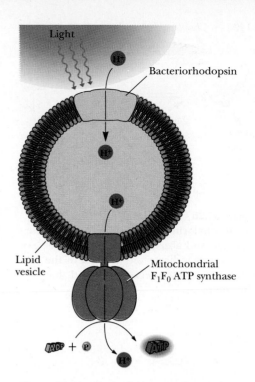

Figure 35.23 Light-driven proton transport drives the synthesis of ATP in *H. halobium*. Walther Stoeckenius and Efraim Racker constructed this artifical membrane system and used it to demonstrate that the proton gradient established by bR can be utilized by an ATP synthase to synthesize ATP. Note that bR is oriented in an outside-in configuration in this lipid vesicle, relative to its normal orientation in cells.

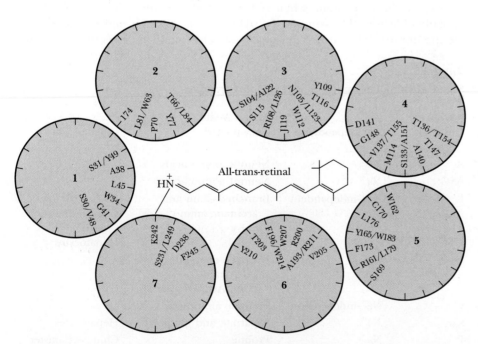

Figure 35.25 A helical wheel model of halorhodopsin. The amino acids facing the polar, hydrophilic core of the protein are shown. Of these 60 residues, 36 are conserved between halorhodopsin and bacteriorhodopsin.

(Adapted from Oesterhelt, D., and Tiltor, J., 1989. Trends in Biochemical Sciences 14:57–61.)

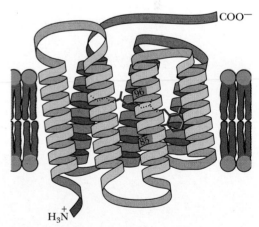

Figure 35.24 The primary sequence of halorhodopsin with the transmembrane segments indicated. The only lysine residue in the protein is Lys^{242}, to which the retinal chromophore is covalently linked.

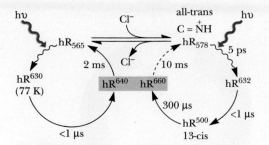

Figure 35.26 The photocycle of light-adapted halorhodopsin (hR), shown in the presence and absence of chloride. The superscripts indicate the maxima of the difference spectra between hR and the intermediates.

transport amino acids and sugars in certain cells. Many of these systems operate as **symports,** with the ion and the transported amino acid or sugar moving in the same direction (that is, into the cell). In **antiport** processes, the ion and the other transported species move in opposite directions. (For example, the anion transporter of erythrocytes is an antiport.) **Proton symport** proteins are used by *E. coli* and other bacteria to accumulate lactose, arabinose, ribose, and a variety of amino acids. *E. coli* also possesses Na^+-symport systems for melibiose as well as for glutamate and other amino acids.

Table 35.3 lists several systems that transport amino acids into mammalian cells. The accumulation of neutral amino acids in the liver by System A represents an important metabolic process. Thus, plasma membrane transport of alanine is the rate-limiting step in hepatic alanine metabolism. This system is normally expressed at low levels in the liver, but substrate deprivation and hormonal activation both stimulate System A expression. Induction of transport activity by either means is inhibited by actinomycin D and α-amanitin, consistent with transcriptional control of the corresponding gene.

Lactose Permease Actively Transports Lactose into *E. coli*

Perhaps the best-understood ion-substrate symport process is the **lactose permease** (also called **lactose/H^+ symport**), which actively transports lactose into *E. coli* cells, accompanied by one H^+ for each lactose transported. This pump derives the energy needed for lactose transport from the **proton-motive force** across the bacterial membrane. Since the proton-motive force is the sum of terms for the proton gradient and the electrical potential difference across the membrane (Chapter 20),

$$\Delta G = RT \ln \frac{[H^+_{in}]}{[H^+_{out}]} + n\mathcal{F}\Delta\psi$$

either a proton gradient *or* an electrical potential difference or both could be used to transport lactose into *E. coli*. Both of these may exist across the *E. coli* membrane as the result of normal respiratory activity (Figure 35.27). The lactose permease protein, when reconstituted into phospholipid vesicles, is capable of lactose/H^+ symport if a pH gradient is established across the vesicle membrane. The lactose permease protein is a 47-kD integral membrane

Table **35.3**
Some Mammalian Amino Acid Transport Systems

System Designation	Ion Dependence	Amino Acids Transported	Cellular Source
A	Na^+	Neutral amino acids	
ASC	Na^+	Neutral amino acids	
L	Na^+-independent	Branched-chain and aromatic amino acids	Ehrlich ascites cells Chinese hamster ovary cells Hepatocytes
N	Na^+	Nitrogen-containing side chains (Gln, Asn, His, etc.)	
y^+	Na^+-independent	Cationic amino acids	
x^-_{AG}	Na^+	Aspartate and glutamate	Hepatocytes
P	Na^+	Proline	Chinese hamster ovary cells

Adapted from: Collarini, E. J., and Oxender, D. L., 1987. Mechanisms of transport of amino acids across membranes. *Annual Review of Nutrition* 7:75–90.

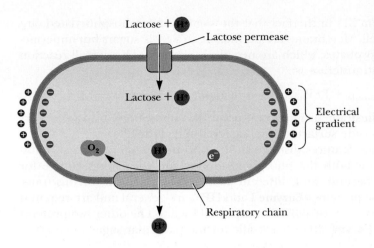

***Figure* 35.27**   The lactose permease in the *E. coli* membrane. An electrical potential difference across the membrane is indicated (negative inside). The flow of electrons through the respiratory chain generates a proton gradient, which is utilized by the lactose permease to transport lactose into the cell.

protein (417 residues) encoded by the *lacY* gene, the second structural gene of the *lac* operon (Chapter 30). Hydropathy analysis of the primary sequence predicts the existence of 12 hydrophobic, α-helical, transmembrane segments in this protein. The model derived from these studies (Figure 35.28) includes hydrophilic, cytoplasmic N- and C-terminal segments, as well as hydrophilic sequences connecting many of the putative transmembrane helices. The binding site for lactose and other galactosides occurs in a portion of the protein lying deep in the lipid bilayer. His^{322} and Glu^{325} of the permease both undergo protonation and deprotonation during lactose transport. These residues may be binding sites for the transported protons in the symport mechanism.

35.7 Group Translocation

Certain bacteria possess a novel and versatile system for the inward transport of certain sugars. In this process, the sugar becomes phosphorylated during its transport across the membrane; that is, transport and phosphorylation are tightly coupled. This type of process, in which a chemical modification accompanies transport has been denoted **group translocation.** Several such systems are known, but the best understood is the **phosphoenolpyruvate:glycose phosphotransferase system,** or simply the **phosphotransferase system** (or **PTS**), discovered by Saul Roseman of Johns Hopkins University in 1964. The advan-

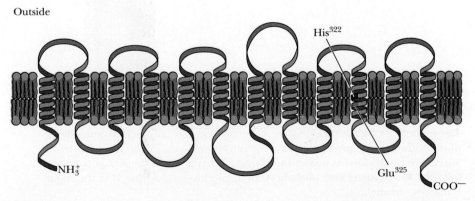

Outside

Inside

***Figure* 35.28**   A model for the arrangement of the lactose permease in the *E. coli* membrane, based on the hydropathy profile of the protein. There are 12 postulated transmembrane segments, and the amino terminus and carboxy terminus are both on the cytoplasmic side of the membrane.

tage of this system lies in the fact that the sugars, once phosphorylated, are trapped in the cell. Membranes are permeable to simple sugars but impermeable to sugar phosphates, which are negatively charged. The overall reaction for the phosphotransferase is:

$$\text{Sugar}_{outside} + \text{PEP}_{inside} \longrightarrow \text{sugar-P}_{inside} + \text{pyruvate}_{inside}$$

The subscripts illustrate an important point: the phosphoryl transfer occurs entirely on the inside surface of the bacterial membrane.

Several unique features distinguish the phosphotransferase. First, phosphoenolpyruvate is both the phosphoryl donor and the energy source for sugar transport. Second, four different proteins are required for this transport. Two of these proteins (**Enzyme I** and **HPr**) are general and are required for the phosphorylation of all PTS-transported sugars. The other two proteins (**Enzyme II** and **Enzyme III**) are specific for the particular sugar to be transported.

The first step in the phosphotransferase reaction (Figure 35.29) is the phosphorylation of Enzyme I by PEP to form a reactive phosphohistidine intermediate (Figure 35.30). This is followed by phosphoryl transfer to a histidyl residue of HPr, followed by phosphorylation of Enzyme III. At the same time, the sugar to be transported is bound on the outside surface of the cell by Enzyme II, which constitutes the sugar transport channel. As the sugar is moved to the inside surface of the membrane, the phosphoryl group is transferred from Enzyme III to the sugar, forming the desired sugar phosphate, which is released into the cytoplasm. (In some cases, for example the *E. coli* mannitol system, no Enzyme III has been identified. In these cases, the C-terminal end of the relevant Enzyme II, which resembles an Enzyme III-type sequence, substitutes for Enzyme III.)

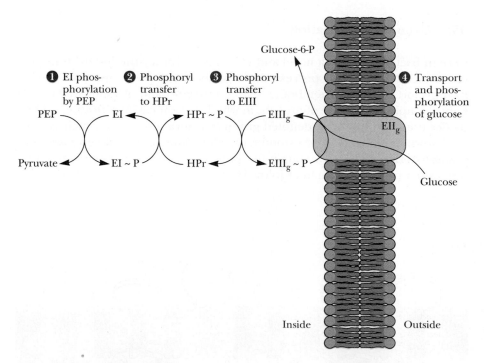

Figure 35.29   Glucose transport in *E. coli* is mediated by the PEP-dependent phosphotransferase system. Enzyme I is phosphorylated in the first step by PEP. Successive phosphoryl transfers to HPr and Enzyme III in Steps 2 and 3 are followed by transport and phosphorylation of glucose. Enzyme II is the sugar transport channel.

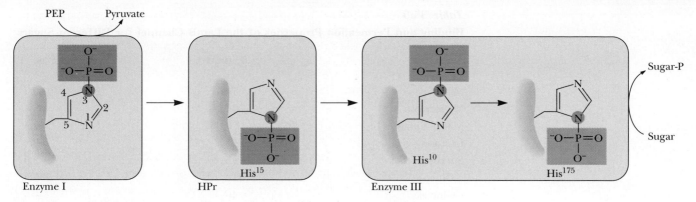

Figure 35.30 The path of the phosphoryl group through the PTS mechanism. Reactive phosphohistidine intermediates of Enzyme I, HPr, and Enzyme III transfer phosphoryl groups from PEP to the transported sugar.

35.8 Specialized Membrane Pores

Porins in Gram-Negative Bacterial Membranes

The membrane transport systems described previously (and many others like them) are relatively specific and function to transport either a single substrate or a very limited range of substrates under normal conditions. At the same time, several rather nonspecific systems also carry out transport processes. One such class of nonspecific transport proteins is found in the outer membranes of Gram-negative bacteria and mitochondria. Low-molecular-weight nutrients and certain other molecules, such as some antibiotics, cross this outer membrane, but larger molecules such as proteins cannot. The ability of the outer membrane to act as a molecular sieve is due to proteins called **porins.** Alternatively, these molecules have been referred to as **peptidoglycan-associated proteins** or simply **matrix proteins. General porins** form nonspecific pores across the outer membrane and sort molecules according to molecular size, whereas **specific porins** contain binding sites for particular substrates. Porins from several organisms have been isolated and characterized (Table 35.4). Molecular masses of the porins generally range from 30 kD to

Table **35.4**

Properties of Some General Porins

Porin and Bacterial Source	Pore Diameter (nm)	M_r Exclusion Limit
E. coli		
OmpF	1.2	
OmpC	1.1	600
PhoE	1.2	
S. typhimurium		
M_r 38,000	1.4	
M_r 39,000	1.4	700
M_r 40,000	1.4	
P. aeruginosa		
F	2.2	6000

Source: Adapted from Benz, R., 1984. Structure and selectivity of porin channels. *Current Topics in Membrane Transport* **21:**199–219, and Benz, R., 1988. Structure and function of porins from Gram-negative bacteria. *Annual Review of Microbiology* **42:**359–393.

Table 35.5

Binding and Permeation Properties of the LamB Channel for Different Sugars

Sugar	K_s (mM)*	P (s^{-1})†
Maltose	10	100
Maltotriose	0.40	66
Maltoheptaose	0.067	2.5
Lactose	56	9
Sucrose	15	2.5
D-Glucose	110	290
L-Glucose	46	—
D-Galactose	42	225
D-Fructose	600	135
D-Mannose	160	160
Stachyose	50	<1

*Half-saturation constant.

†Rate of permeation relative to that of maltose. Data adjusted to 100 s^{-1} for maltose. The LamB-containing liposomes were added to buffer solutions containing 40 mM of the corresponding test sugars.

Source: Adapted from Benz, R. 1988. Structure and function of porins from Gram-negative bacteria. *Annual Review of Microbiology* **42**:359–393.

50 kD. Most (but not all) porins are arranged in the outer membrane as trimers of identical subunits. The molecular exclusion limits clearly depend on the size of the pore formed by the porin molecule. The pores formed by *E. coli* and *S. typhimurium* porins are relatively small, but porin F from *Pseudomonas aeruginosa* creates a much larger pore, with an exclusion limit of approximately 6 kD. Specific porins *LamB* and *Tsx* of *E. coli* and porins *P* and *D1* of *P. aeruginosa* possess specific binding sites for maltose and related oligosaccharides (Table 35.5), nucleosides, anions, and glucose, respectively.

Porins show high degrees of sequence homology and similarity. The most intriguing feature of porin secondary and tertiary structure is this: In contrast to nearly all other membrane proteins that adopt α-helical structures in the transmembrane segments, porins show little or no evidence of α-helical domains and segments. Instead, the porins and other outer membrane proteins adopt β-sheet structures for their membrane-spanning segments. Models of membrane insertion, which involve β-strands arranged perpendicular to the membrane plane, have been proposed for several porins (Figure 35.31). The

Figure **35.31** A model for the arrangement of the porin *PhoE* in the outer membrane of *E. coli*. The transmembrane segments are strands of β-sheet.

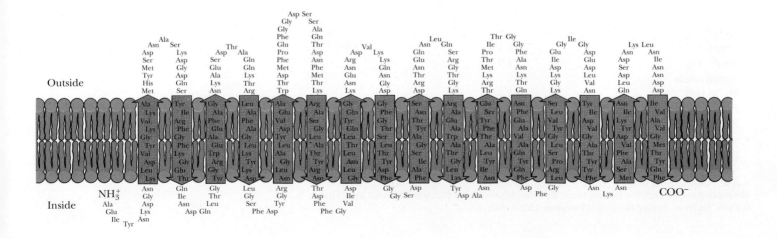

(a)

(b)

(c)

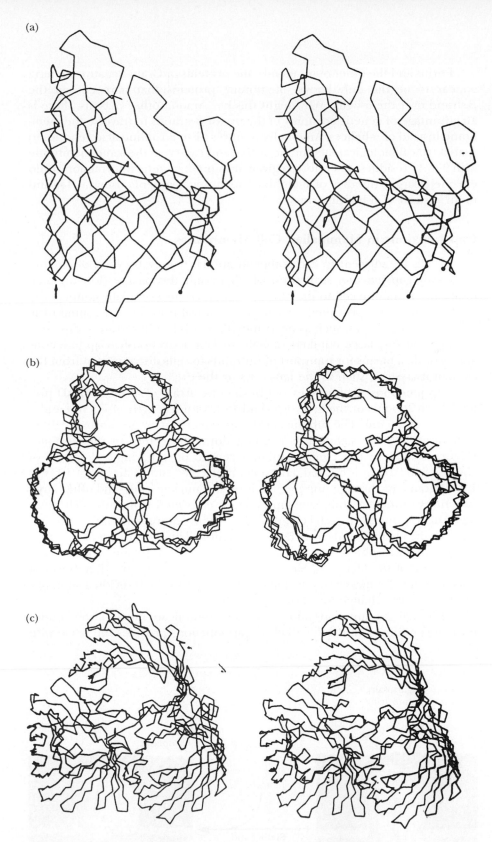

Figure 35.32 Three-dimensional reconstruction of porin from *Rhodobacter capsulatus.*

crystal structure of the porin from *Rhodobacter capsulatus* shows a trimer in which each monomer forms a pore (Figure 35.32). The monomer pore consists of a 16-stranded β-barrel that traverses the membrane as a tube. The tube is narrowed near the center by peptide chain segments protruding from the inner wall of the barrel. These chain segments form an "eyelet" about 1 nm long and 0.6 to 1 nm across. The eyelet is postulated to determine the exclusion limit for particles diffusing through the pore.

Porins and the other outer membrane proteins of Gram-negative bacteria appear to be the only known membrane proteins that have chosen the β-strand over the α-helix. Why might this be? Among other reasons, there is an advantage of genetic economy in the use of β-strands to traverse the membrane instead of α-helices. An α-helix requires 21 to 25 amino acid residues to span a typical biological membrane; a β-strand can cross the same membrane with 9 to 11 residues. Therefore, a given amount of genetic information could encode a larger number of membrane-spanning segments using a β-strand motif instead of α-helical arrays.

Gap Junctions in Mammalian Cell Membranes

When cells lie adjacent to each other in animal tissues, they are often connected by **gap junction** structures, which permit the passive flow of small molecules from one cell to the other. Such junctions essentially connect the cells metabolically, providing a means of chemical transfer and communication. In certain tissues, such as heart muscle that is not innervated, gap junctions permit very large numbers of cells to act synchronously. Gap junctions also provide a means for transport of nutrients to cells disconnected from the circulatory system, such as the lens cells of the eye.

Gap junctions are formed from hexameric arrays of a single 32-kD protein. Each subunit of the array is cylindrical, with a length of 7.5 nm and a diameter of 2.5 nm. The subunits of the hexameric array are normally tilted with respect to the sixfold axis running down the center of the hexamer (Figure 35.33). In this conformation, a central pore having a diameter of about 1.8 to 2.0 nm is created, and small molecules (up to masses of 1 kD to 1.2 kD) can pass through unimpeded. Proteins, nucleic acids, and other large structures cannot. A complete gap junction is formed from two such hexameric arrays, one from each cell. A twisting, sliding movement of the subunits narrows the channel and closes the gap junction. This closure is a cooperative process, and a localized conformation change at the cytoplasmic end assists in the closing of the channels. Since the closing of the gap junction does not appear to involve massive conformational changes in the individual subunits, the free energy change for closure is small.

Although gap junctions allow cells to communicate metabolically under normal conditions, the ability to close gap junctions provides the tissue with

Figure **35.33** Gap junctions consist of hexameric arrays of cylindrical protein subunits in the plasma membrane. The subunit cylinders are tilted with respect to the axis running through the center of the gap junction. A gap junction between cells is formed when two hexameric arrays of subunits in separate cells contact each other and form a pore through which cellular contents may pass. Gap junctions close by means of a twisting, sliding motion in which the subunits decrease their tilt with respect to the central axis. Closure of the gap junction is Ca^{2+}-dependent.

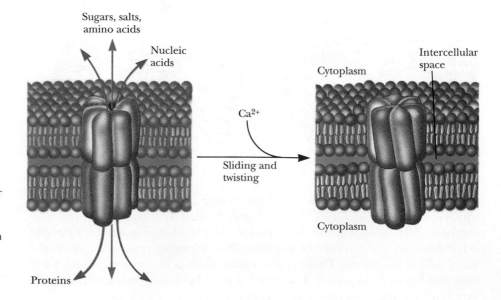

an important intercellular regulation mechanism. In addition, gap junctions provide a means to protect adjacent cells if one or more cells are damaged or stressed. To these ends, gap junctions are sensitive to membrane potentials, hormonal signals, pH changes, and intracellular calcium levels. Dramatic changes in pH or Ca^{2+} concentration in a cell may be a sign of cellular damage or death. In order to protect neighboring cells from the propagation of such effects, gap junctions close in response to decreased pH or prolonged increases in intracellular Ca^{2+}. Under normal conditions of intracellular Ca^{2+} levels ($<10^{-7}$ *M*), gap junctions are open and intercellular communication is maintained. When calcium levels rise to 10^{-5} *M* or higher, the junctions, sensing danger, rapidly close. The Hill coefficient (Chapter 12) for H^{+}-stimulated closing is about 4.5, whereas that for Ca^{2+} is approximately 3, consistent with the idea that the closing of gap junctions is a cooperative process.

35.9 Ionophore Antibiotics

All of the transport systems examined thus far are relatively large proteins. Several small molecule toxins produced by microorganisms facilitate ion transport across membranes. Due to their relative simplicity, these molecules, the **ionophore antibiotics,** represent paradigms of the **mobile carrier** and **pore** or **channel** models for membrane transport. Mobile carriers are molecules that form complexes with particular ions and diffuse freely across a lipid membrane (Figure 35.34). Pores or channels, on the other hand, adopt a fixed orientation in a membrane, creating a hole that permits the transmembrane movement of ions. These pores or channels may be formed from monomeric or (more often) multimeric structures in the membrane.

Carriers and channels may be distinguished on the basis of their temperature dependence. Channels are comparatively insensitive to membrane phase transitions and show only a slight dependence of transport rate on temperature. Mobile carriers, on the other hand, function efficiently above a membrane phase transition, but only poorly below it. Consequently, mobile carrier systems often show dramatic increases in transport rate as the system is heated through its phase transition. Figure 35.35 displays the structures of several of

(a) Carrier ionophore

(b) Channel-forming ionophore

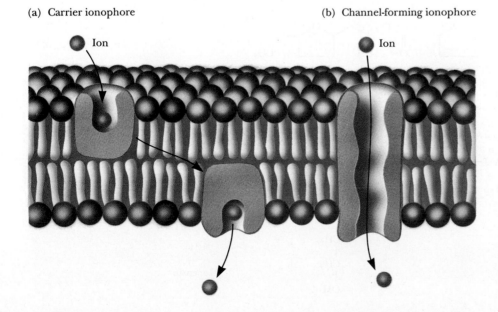

Ion

Ion

Figure **35.34**   Schematic drawings of mobile carrier and channel ionophores. Carrier ionophores must move from one side of the membrane to the other, acquiring the transported species on one side and releasing it on the other side. Channel ionophores span the entire membrane.

these interesting molecules. As might be anticipated from the variety of structures represented here, these molecules associate with membranes and facilitate transport by different means.

Valinomycin Is a Mobile Carrier Ionophore

Valinomycin (isolated from *Streptomyces fulvissimus*) is a cyclic structure containing 12 units made from 4 different residues. Two are amino acids (L-valine and D-valine); the other two residues, L-lactate and D-hydroxyisovalerate, contribute ester linkages. Valinomycin is a **depsipeptide,** that is, a molecule with

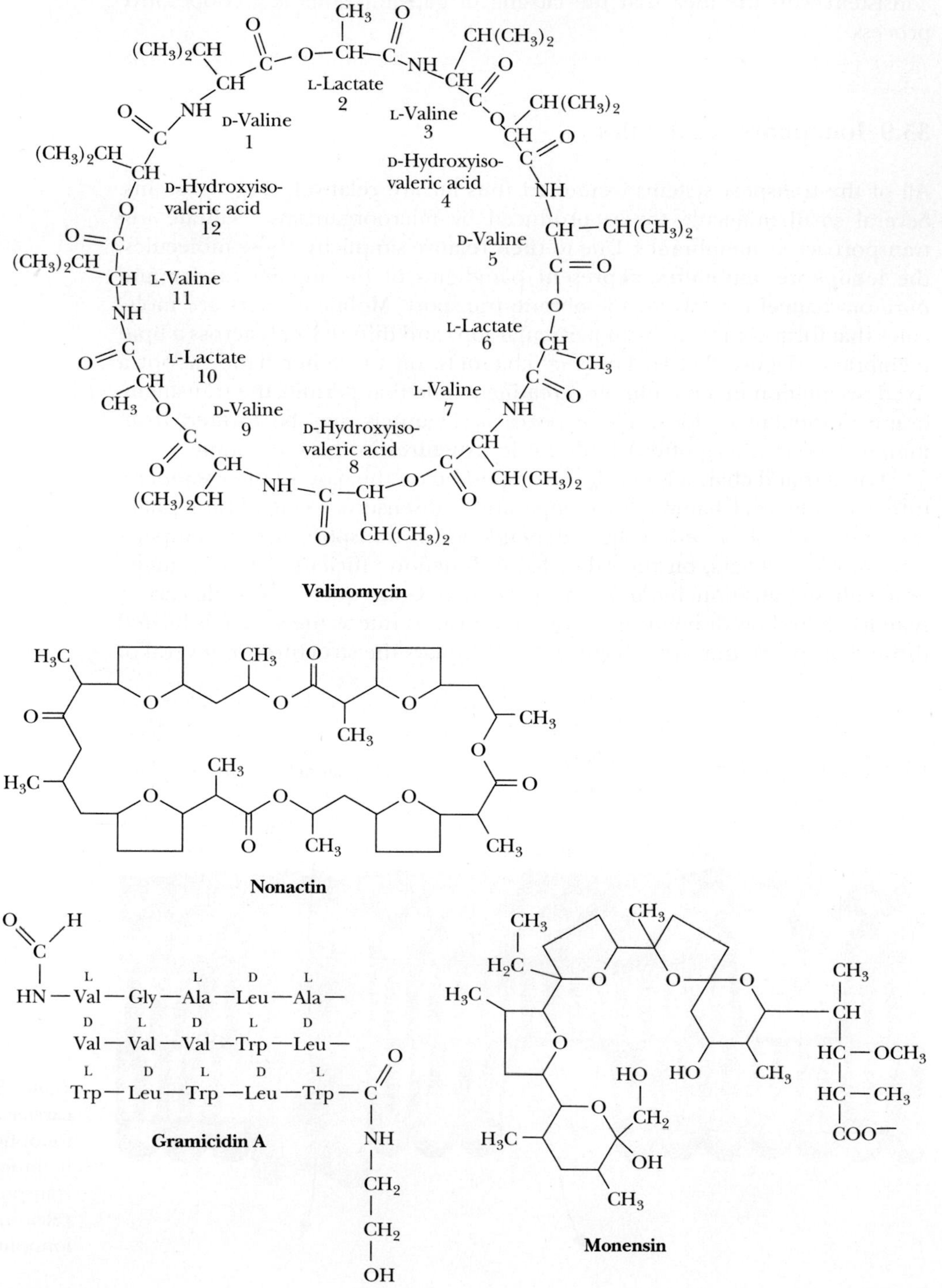

Valinomycin

Nonactin

Gramicidin A

Monensin

***Figure* 35.35** Structures of several ionophore antibiotics. Valinomycin consists of three repeats of a four-unit sequence. Because it contains both peptide and ester bonds, it is referred to as a depsipeptide.

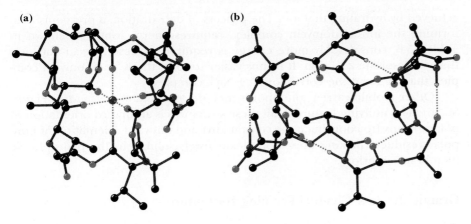

(a)   **(b)**

Figure **35.36**   The structures of the valinomycin-K^+ complex and of uncomplexed valinomycin.

both peptide and ester bonds. (Considering the 12 units in the structure, valinomycin is called a **dodecadepsipeptide.**) Valinomycin consists of the 4-unit sequence (D-valine, L-lactate, L-valine, D-hydroxyisovaleric acid), repeated three times to form the cyclic structure in Figure 35.35. The structures of uncomplexed valinomycin and the K^+-valinomycin complex have been studied by X-ray crystallography (Figure 35.36). The structure places K^+ at the center of the valinomycin ring, coordinated with the carbonyl oxygens of the 6 valines. The polar groups of the valinomycin structure are positioned toward the center of the ring, whereas the nonpolar groups (the methyl and isopropyl side chains) are directed outward from the ring. The hydrophobic exterior of valinomycin interacts favorably with low dielectric solvents and with the hydrophobic interiors of lipid bilayers. Moreover, the central carbonyl groups completely surround the K^+ ion, shielding it from contact with nonpolar solvents or the hydrophobic membrane interior. As a result, the K^+-valinomycin complex freely diffuses across biological membranes and effects rapid, passive K^+ transport (up to 10,000 K^+/sec) in the presence of K^+ gradients.

Valinomycin displays a striking selectivity with respect to monovalent cation binding. It binds K^+ and Rb^+ tightly, but shows about a thousandfold lower affinity for Na^+ and Li^+. The smaller ionic radii of Na^+ and Li^+ (compared to K^+ and Rb^+) may be responsible in part for the observed differences. However, another important difference between Na^+ and K^+ is shown in Table 35.6. The **free energy of hydration** for an ion is the stabilization

Table **35.6**

Properties of Alkali Cations

Ion	Atomic Number	Ionic Radius (nm)	Hydration Free Energy, ΔG (kJ/mol)
Li^+	3	0.06	−410
Na^+	11	0.095	−300
K^+	19	0.133	−230
Rb^+	37	0.148	−210
Cs^+	55	0.169	−200

achieved by hydrating that ion. The process of dehydration, a prerequisite to forming the ion-valinomycin complex, requires energy input. As shown in Table 35.6, considerably more energy is required to desolvate an Na^+ ion than to desolvate a K^+ ion. It is thus easier to form the K^+-valinomycin complex than to form the corresponding Na^+ complex.

Other mobile carrier ionophores include *monensin* and *nonactin* (Figure 35.36). The unifying feature in all these structures is an inward orientation of polar groups (to coordinate the central ion) and outward orientation of nonpolar residues (making these complexes freely soluble in the hydrophobic membrane interior).

Gramicidin Is a Channel-Forming Ionophore

In contrast to valinomycin, all naturally occurring membrane transport systems appear to function as channels, not mobile carriers. All of the proteins discussed in this chapter use multiple transmembrane segments to create channels in the membrane, through which species are transported. For this reason, it may be more relevant to consider the **pore** or **channel** ionophores. **Gramicidin** from *Bacillus brevis* (Figure 35.35) is a linear peptide of 15 residues and is a prototypical channel ionophore. Gramicidin contains alternating L- and D-residues, a formyl group at the N-terminus, and an ethanolamine at the C-terminus. The predominance of hydrophobic residues in the gramicidin structure facilitates its incorporation into lipid bilayers and membranes. Once incorporated in lipid bilayers, it permits the rapid diffusion of many different cations. Gramicidin possesses considerably less ionic specificity than does valinomycin, but permits higher transport rates. A single gramicidin channel can transport as many as 10 million K^+ ions per second. Protons and all alkali cations can diffuse through gramicidin channels, but divalent cations such as Ca^{2+} block the channel.

Gramicidin forms two different helical structures. A double helical structure predominates in organic solvents (Figure 35.37), whereas a helical dimer is formed in lipid membranes. (An α-helix cannot be formed by gramicidin, because it has both D- and L-amino acid residues.) The helical dimer is a head-to-head or amino terminus-to-amino terminus (N-to-N) dimer oriented perpendicular to the membrane surface, with the formyl groups at the bilayer center and the ethanolamine moieties at the membrane surface. The helix is unusual, with 6.3 residues per turn and a central hole approximately 0.4 nm in diameter. The hydrogen-bonding pattern in this structure, in which N-H groups alternately point up and down the axis of the helix to hydrogen-bond with carbonyl groups, is reminiscent of a β-sheet. For this reason this structure has often been referred to as a β-helix.

Amphipathic Helices Form Transmembrane Ion Channels

Recently, a variety of natural peptides that form transmembrane channels have been identified and characterized. Melittin (Figure 35.38) is a bee venom toxin peptide of 26 residues. The cecropins are peptides induced in *Hyalophora cecropia* (Figure 35.39) and other related silkworms when challenged by bacterial infections. These peptides are thought to form α-helical aggregates in membranes, creating an ion channel in the center of the aggregate. The unifying feature of these helices is their **amphipathic** character, with polar residues clustered on one face of the helix and nonpolar residues elsewhere. In the membrane, the polar residues face the ion channel, leaving the nonpolar residues elsewhere on the helix to interact with the hydrophobic interior of the lipid bilayer.

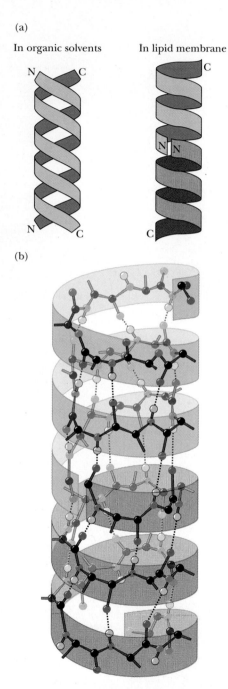

(a)

In organic solvents   In lipid membrane

(b)

***Figure* 35.37**   (a) Gramicidin forms a double helix in organic solvents; a helical dimer is the preferred structure in lipid bilayers. The structure is a head-to-head, left-handed helix, with the carboxy termini of the two monomers at the ends of the structure. (b) The hydrogen-bonding pattern resembles that of a parallel β-sheet.

Alamethicin I[1]:

Ac-Aib-Pro-Aib-Ala-Aib-Ala-Gln-Aib-Val-Aib-Gly-Leu-Aib-Pro-Val-Aib-Aib-Glu-Gln-Phol*

Cecropin A:

Lys-Trp-Lys-Leu-Phe-Lys-Lys-Ile-Glu-Lys-Val-Gly-Gln-Asn-Ile-Arg-Asp-Gly-Ile-Ile-Lys-Ala-Gly-Pro-Ala-Val-Ala-Val-Val-Gly-Gln-Ala-Thr-Gln-Ile-Ala-Lys-NH₂

Melittin:

Gly-Ile-Gly-Ala-Val-Leu-Lys-Val-Leu-Thr-Thr-Gly-Leu-Pro-Ala-Leu-Ile-Ser-Trp-Ile-Lys-Arg-Lys-Arg-Gln-Gln-NH₂

Magainin 2 amide:

Gly-Ile-Gly-Lys-Phe-Leu-His-Ser-Ala-Lys-Lys-Phe-Gly-Lys-Ala-Phe-Val-Gly-Glu-Ile-Met-Asn-Ser-NH₂

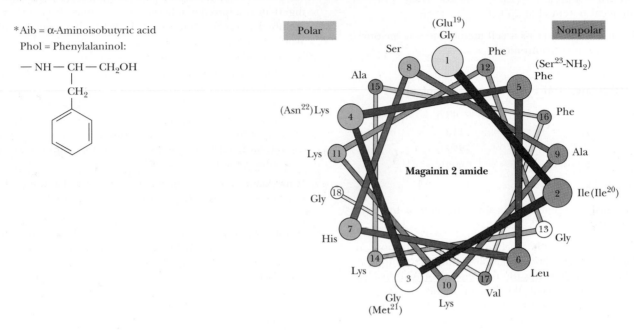

Figure 35.38 The structures of several amphipathic peptide antibiotics. α-Helices formed from these peptides cluster polar residues on one face of the helix, with nonpolar residues at other positions.

Figure 35.39 (*left*) Adult and (*right*) caterpillar stages of the cecropia moth, *Hyalophora cecropia.*

Problems

1. Calculate the free energy difference at 25°C due to a galactose gradient across a membrane, if the concentration on side 1 is 2 mM and the concentration on side 2 is 10 mM.

2. Consider a phospholipid vesicle containing 10 mM Na$^+$ ions. The vesicle is bathed in a solution that contains 52 mM Na$^+$ ions, and the electrical potential difference across the vesicle membrane $\Delta\psi = \psi_{outside} - \psi_{inside} = -30$ mV. What is the electrochemical potential at 25°C for Na$^+$ ions?

3. Transport of histidine across a cell membrane was measured at several histidine concentrations:

[Histidine], μM	Transport, $\mu mol/min$
2.5	42.5
7	119
16	272
31	527
72	1220

Does this transport operate by passive diffusion or by facilitated diffusion?

4. Fructose is present outside a cell at 1 μM concentration. An active transport system in the plasma membrane transports fructose into this cell, using the free energy of ATP hydrolysis to drive fructose uptake. What is the highest intracellular concentration of fructose that this transport system can generate? Assume that one fructose is transported per ATP hydrolyzed, that ATP is hydrolyzed on the intracellular surface of the membrane, and that the concentrations of ATP, ADP, and P$_i$ are 3 mM, 1 mM, and 0.5 mM, respectively. T = 298 K. (*Hint:* Refer to Chapter 16 to recall the effects of concentration on free energy of ATP hydrolysis.)

5. The rate of K$^+$ transport across bilayer membranes reconstituted from dipalmitoylphosphatidylcholine (DPPC) and nigericin is approximately the same as that observed across membranes reconstituted from DPPC and *cecropin a* at 35°C. Would you expect the transport rates across these two membranes to also be similar at 50°C? Explain.

6. Fluoride ion inhibits the uptake of glucose and similar sugars by the phosphotransferase system (PTS) in *E. coli.* However, no effect of fluoride that would explain this inhibition has been observed with any of the proteins of the PTS system. Fluoride ion has, however, long been known to inhibit the glycolytic enzyme, enolase. Can you explain these puzzling observations?

7. It has been observed that dicumarol and FCCP inhibit lactose transport across the plasma membrane of *E. coli.* Explain.

8. In this chapter, we have examined coupled transport systems that rely on ATP hydrolysis, on primary gradients of Na$^+$ or H$^+$, and on phosphotransferase systems. Suppose you have just discovered an unusual strain of bacteria that transports rhamnose across its plasma membrane. Suggest experiments that would test whether it was linked to any of these other transport systems.

Further Reading

Benz, R., 1980. Structure and function of porins from Gram-negative bacteria. *Annual Review of Microbiology* **42:**359–393.

Blair, H. C., et al., 1989. Osteoclastic bone resorption by a polarized vacuolar proton pump. *Science* **245:**855–857.

Christensen, B., et al., 1988. Channel-forming properties of cecropins and related model compounds incorporated into planar lipid membranes. *Proceedings of the National Academy of Sciences, U.S.A.* **85:**5072–5076.

Collarini, E. J., and Oxender, D., 1987. Mechanisms of transport of amino acids across membrane. *Annual Review of Nutrition* **7:**75–90.

Featherstone, C., 1990. An ATP-driven pump for secretion of yeast mating factor. *Trends in Biochemical Sciences* **15:**169–170.

Garavito, R. M., et al., 1983. X-ray diffraction analysis of matrix porin, an integral membrane protein from *Escherichia coli* outer membrane. *Journal of Molecular Biology* **164:**313–327.

Glynn, I., and Karlish, S., 1990. Occluded ions in active transport. *Annual Review of Biochemistry* **59:**171–205.

Gould, G. W., and Bell, G. I., 1990. Facilitative glucose transporters: An expanding family. *Trends in Biochemical Sciences* **15:**18–23.

Inesi, G., Sumbilla, C., and Kirtley, M., 1990. Relationships of molecular structure and function in Ca^{2+}-transport ATPase. *Physiological Reviews* **70:**749–759.

Jap, B., and Walian, P. J., 1990. Biophysics of the structure and function of porins. *Quarterly Reviews of Biophysics* **23:**367–403.

Jay, D., and Cantley, L., 1986. Structural aspects of the red cell anion exchange protein. *Annual Review of Biochemistry* **55:**511–538.

Jennings, M. L., 1989. Structure and function of the red blood cell anion transport protein. *Annual Review of Biophysics and Biophysical Chemistry* **18:**397–430.

Jørgensen, P. L., 1986. Structure, function and regulation of Na$^+$,K$^+$-ATPase in the kidney. *Kidney International* **29:**10–20.

Jørgensen, P. L., and Andersen, J. P., 1988. Structural basis for E$_1$-E$_2$ conformational transitions in Na$^+$,K$^+$-pump and Ca-pump proteins. *Journal of Membrane Biology* **103:**95–120.

Juranka, P. F., Zastawny, R. L., and Ling, V., 1989. P-Glycoprotein: Multidrug-resistance and a superfamily of membrane-associated transport proteins. *The FASEB Journal* **3:**2583–2592.

Kaback, H. R., Bibi, E., and Roepe, P. D., 1990. β-Galactoside transport in *E. coli:* A functional dissection of *lac* permease. *Trends in Biochemical Sciences* **15:**309–314.

Kartner, N., and Ling, V., 1989. Multidrug resistance in cancer. *Scientific American* March **260**:44–51.

Matthew, M. K., and Balaram, A., 1983. A helix dipole model for alamethicin and related transmembrane channels. *FEBS Letters* **157**:1–5.

Meadow, N. D., Fox, D. K., and Roseman, S., 1990. The bacterial phosphoenolpyruvate:glycose phosphotransferase system. *Annual Review of Biochemistry* **59**:497–542.

Oesterhelt, D., and Tittor, J., 1989. Two pumps, one principle: Light-driven ion transport in *Halobacteria. Trends in Biochemical Sciences* **14**:57–61.

Pedersen, P. L., and Carafoli, E., 1987. Ion motive ATPases. *Trends in Biochemical Sciences* **12**:146–150, 186–189.

Pressman, B., 1976. Biological applications of ionophores. *Annual Review of Biochemistry* **45**:501–530.

Spudich, J. L., and Bogomolni, R. A., 1988. Sensory rhodopsins of *Halobacteria. Annual Review of Biophysics and Biophysical Chemistry* **17**:193–215.

Wade, D., et al., 1990. All-ᴅ amino acid-containing channel-forming antibiotic peptides. *Proceedings of the National Academy of Sciences, U.S.A.* **87**:4761–4765.

Wallace B. A., 1990. Gramicidin channels and pores. *Annual Review of Biophysics and Biophysical Chemistry* **19**:127–157.

Walmsley, A. R., 1988. The dynamics of the glucose transporter. *Trends in Biochemical Sciences* **13**:226–231.

Wheeler, T. J., and Hinkle, P., 1985. The glucose transporter of mammalian cells. *Annual Review of Physiology* **47**:503–517.

Chapter 36

Muscle Contraction

Michaelangelo's "David" epitomizes the
musculature of the human form.

Movement is an intrinsic property associated with all living things. Within cells, molecules undergo coordinated and organized movements, and cells themselves may move across a surface. At the tissue level, **muscle contraction** allows higher organisms to carry out and control crucial internal functions, such as peristalsis in the gut and the beating of the heart. Muscle contraction also enables the organism to carry out organized and sophisticated movements, such as walking, running, flying, and swimming.

The movement of muscle tissue bears many similarities to the movement carried out by cilia and flagella and cytoskeletal elements (Chapter 34). The source of energy for muscle movement is ATP. The binding of ATP and its subsequent hydrolysis control conformational changes that result in sliding or walking movements of one molecule relative to another. As fundamental and straightforward as all this sounds, elucidation of these basically simple processes has been extremely challenging for biochemists, involving the application of many sophisticated chemical and physical methods in many different laboratories. This chapter describes the structure and the chemical processes of muscle tissue and some important experiments on muscle contraction.

36.1 The Morphology of Muscle

Four different kinds of muscle are found in animals (Figure 36.1). They are **skeletal** muscle, **cardiac (heart)** muscle, **smooth** muscle, and **myoepithelial cells.** The cells of the latter three types contain only a single nucleus and are called **myocytes.** The cells of skeletal muscle are long and multinucleate and are referred to as **muscle fibers.** At the microscopic level, skeletal muscle and cardiac muscle display alternating light and dark bands, and for this reason are often referred to as **striated** muscles. The different types of muscle cells vary widely in structure, size, and function. In addition, the times required for contractions and relaxations by various muscle types vary considerably. The fastest responses (on the order of milliseconds) are observed for **fast-twitch** skeletal muscle, and the slowest responses (on the order of seconds) are found in smooth muscle. **Slow-twitch** skeletal muscle tissue displays an intermediate response time.

Structural Features of Skeletal Muscle

Skeletal muscles in higher animals consist of 100-μm-diameter **fiber bundles,** some as long as the muscle itself. Each of these muscle fibers contains hundreds of **myofibrils** (Figure 36.2), each of which spans the length of the fiber and is about 1 to 2 μm in diameter. Myofibrils are linear arrays of cylindrical **sarcomeres,** the basic structural units of muscle contraction. The sarcomeres are surrounded on each end by a membrane system that is actually an elaborate extension of the muscle fiber plasma membrane or **sarcolemma.** These extensions of the sarcolemma, which are called **transverse tubules** or **t-tubules,** enable the sarcolemmal membrane to contact the ends of each myofibril in the muscle fiber (Figure 36.2). This important topological feature is crucial to the initiation of contractions. In between the t-tubules, the sarcomere is covered with a specialized endoplasmic reticulum called the **sarcoplasmic reticulum,** or **SR.** The SR contains high concentrations of Ca^{2+}, and the release of Ca^{2+} from the SR and its interactions within the sarcomeres

(a) Part of a skeletal   (b) Heart muscle cells   (c) Smooth muscle cells   (d) Myoepithelial cell
    muscle cell

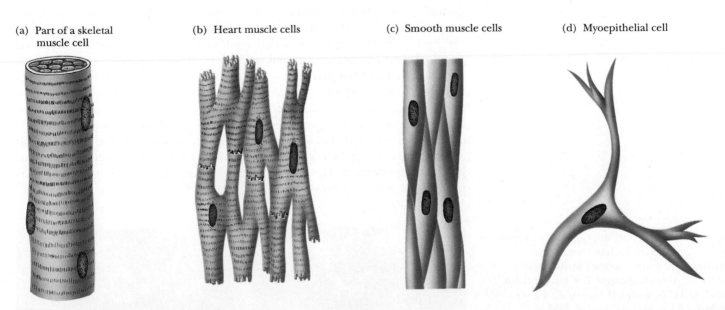

***Figure* 36.1** The four classes of muscle cells in mammals. Skeletal muscle and cardiac muscle are striated. Cardiac muscle, smooth muscle, and myoepithelial cells are mononucleate, whereas skeletal muscle is multinucleate.

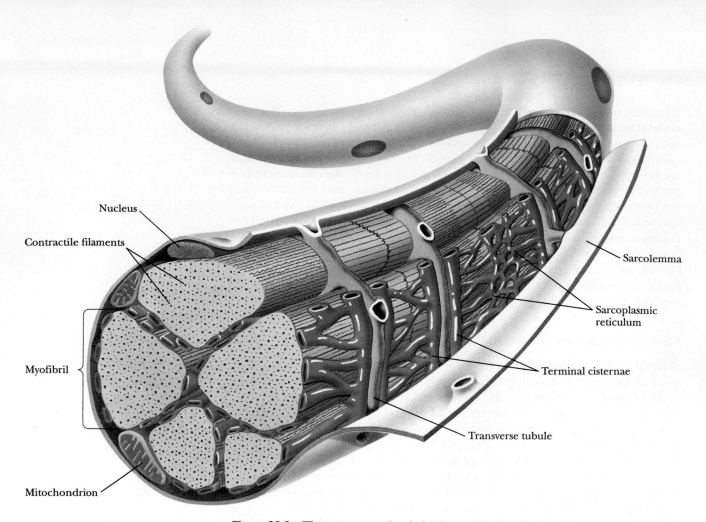

Nucleus

Contractile filaments

Myofibril

Mitochondrion

Sarcolemma

Sarcoplasmic reticulum

Terminal cisternae

Transverse tubule

Figure 36.2   The structure of a skeletal muscle cell, showing the manner in which t-tubules enable the sarcolemmal membrane to contact the ends of each myofibril in the muscle fiber.

trigger muscle contraction, as we will see. Each SR structure consists of two domains. **Longitudinal tubules** run the length of the sarcomere and are capped on either end by the **terminal cisternae** (Figure 36.2). The structure at the end of each sarcomere, which consists of a t-tubule and two apposed terminal cisternae, is called a **triad**, and the intervening gaps of approximately 15 nm are called **triad junctions.** The junctional face of each terminal cisterna is joined to its respective t-tubule by a **foot structure** (Figure 36.3). Skeletal muscle contractions are initiated by nerve stimuli that act directly on the muscle. Nerve impulses produce an electrochemical signal (see Chapter 38) called an action potential that spreads over the sarcolemmal membrane and into the fiber along the t-tubule network. This signal is passed across the triad junction and induces the release of Ca^{2+} ions from the SR. These Ca^{2+} ions bind to the muscle fibers and induce contraction.

Figure 36.3   Electron micrograph of rabbit fast-twitch skeletal muscle membrane, illustrating the interactions at the triad junction. Abbreviations: LT = longitudinal tubules, TC = terminal cisternae, TT = t-tubule, FS = foot structures, JFM = junctional face membrane of SR, CPM = calcium pump membrane.

(From: Fleischer, S., and Inui, M., 1989. Annual Review of Biophysics and Biophysical Chemistry 18:333–366.)

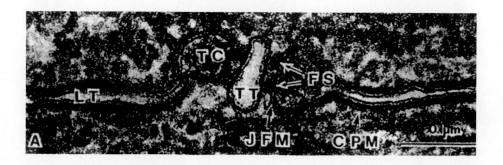

36.2 The Molecular Structure of Skeletal Muscle

Examination of myofibrils in the electron microscope reveals a banded or striated structure. The bands are traditionally identified by letters (Figure 36.4). Regions of high electron density, denoted **A bands,** alternate with regions of low electron density, the **I bands.** Small, dark **Z lines** lie in the middle of the I bands, marking the ends of the sarcomere. Each A band has a central region of slightly lower electron density called the **H zone,** which contains a central **M disk** (also called an **M line**). Electron micrographs of cross-sections of each of these regions reveal molecular details. The H zone shows a regular, hexagonally arranged array of **thick filaments** (15 nm diameter), whereas the I band shows a regular, hexagonal array of **thin filaments** (7 nm diameter). In the dark regions at the ends of each A band, the thin and thick filaments interdigitate, as shown in Figure 36.4. The thin filaments are composed primarily of three proteins called **actin, troponin,** and **tropomyosin.** The thick filaments consist mainly of a protein called **myosin.** The thin and thick filaments are joined by **cross-bridges.** These cross-bridges are actually extensions of the myosin molecules, and muscle contraction is accomplished by the sliding of the cross-bridges along the thin filaments, a mechanical movement driven by the free energy of ATP hydrolysis.

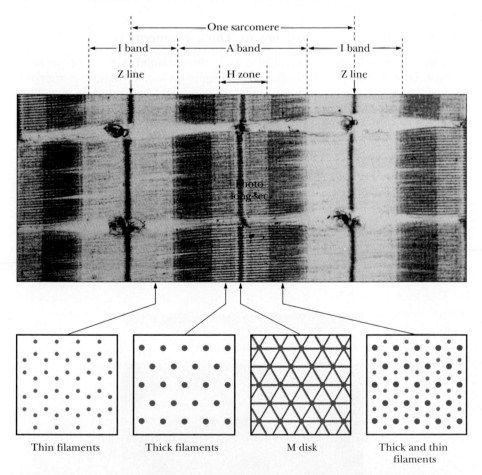

Figure 36.4   Electron micrograph of a skeletal muscle myofibril (in longitudinal section). The length of one sarcomere is indicated, as are the A and I bands, the H zone, the M disk, and the Z lines. Cross-sections from the H zone show a hexagonal array of thick filaments, whereas the I band cross-section shows a hexagonal array of thin filaments.

(Photo courtesy of Hugh Huxley, Brandeis University.)

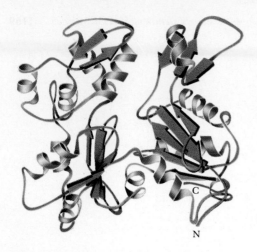

Figure 36.5 The three-dimensional structure of an actin monomer from skeletal muscle. This view shows the two domains (left and right) of actin.

The Composition and Structure of Thin Filaments

Actin, the principal component of thin filaments, can be isolated in two forms. Under conditions of low ionic strength, actin exists as a 42-kD globular protein, denoted **G-actin.** G-actin consists of two principal lobes or domains (Figure 36.5). At physiological conditions (higher ionic strength), G-actin polymerizes to form a *fibrous* form of actin, called **F-actin.** As shown in Figure 36.6, F-actin is a right-handed helical structure, with a helix pitch of about 72 nm per turn. The F-actin helix is the core of the thin filament, to which *tropomyosin* and the **troponin complex** also add. Tropomyosin is a dimer of homologous but nonidentical 33-kD subunits. These two subunits form long α-helices that intertwine, creating 38- to 40-nm-long coiled coils, which join in head-to-tail fashion to form long rods. These rods bind to the F-actin polymer and lie almost parallel to the long axis of the F-actin helix (Figure 36.7a–c). Each tropomyosin heterodimer contacts approximately seven actin subunits. The troponin complex consists of three different proteins: **troponin T,** or **TnT** (37 kD); **troponin I,** or **TnI** (24 kD); and **troponin C,** or **TnC** (18 kD). TnT binds to tropomyosin, specifically at the head-to-tail junction. Troponin I binds both to tropomyosin and to actin. Troponin C is a Ca^{2+}-binding protein that binds to TnI. TnC shows 70% homology with the important Ca^{2+} signaling protein, calmodulin (Chapter 37). The release of Ca^{2+} from the SR, which signals a contraction, raises the cytosolic Ca^{2+} concentration high enough to saturate the Ca^{2+} sites on TnC. Ca^{2+} binding induces a conformational change in the amino-terminal domain of TnC, which in turn causes a rearrangement of the troponin complex and tropomyosin with respect to the actin fiber.

The Composition and Structure of Thick Filaments

Myosin, the principal component of muscle thick filaments, is a large protein consisting of six polypeptides, with an aggregate molecular weight of approximately 540 kD. As shown in Figure 36.8, the six peptides include two 230-kD **heavy chains,** as well as two pairs of different 20-kD **light chains,** denoted **LC1** and **LC2.** The heavy chains consist of globular amino-terminal **myosin heads,** joined to long α-helical carboxy-terminal segments, the **tails.** These tails are intertwined to form a left-handed coiled coil approximately 2 nm in diameter and 130 to 150 nm long. Each of the heads in this dimeric structure is associated with an LC1 and an LC2. The myosin heads exhibit **ATPase activity,** and hydrolysis of ATP by the myosin heads drives muscle contraction. LC1 is also known as the **essential light chain,** and LC2 is designated the **regulatory light chain.** Both light chains are homologous to calmodulin and TnC. Dissociation of LC1 from the myosin heads by alkali cations results in loss of the myosin ATPase activity.

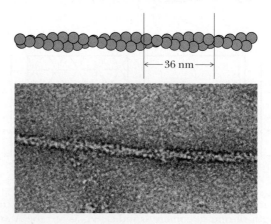

Figure 36.6 The helical arrangement of actin monomers in F-actin. The F-actin helix has a pitch of 72 nm and a repeat distance of 36 nm.
(Electron micrograph courtesy of Hugh Huxley, Brandeis University.)

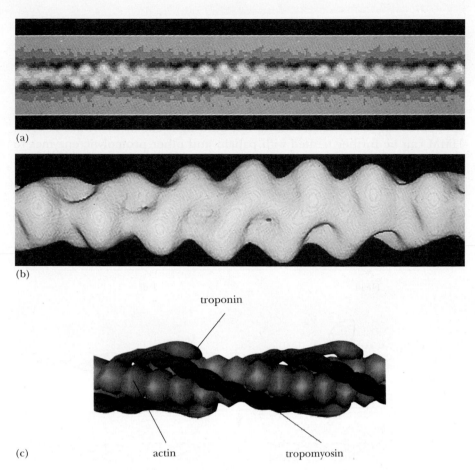

(a)

(b)

troponin

actin   tropomyosin

(c)

***Figure* 36.7**   (a) An electron micrograph of a thin filament, (b) a corresponding image reconstruction, and (c) a schematic drawing based on the images in (a) and (b). The tropomyosin coiled coil winds around the actin helix, each tropomyosin dimer interacting with seven consecutive actin monomers. Troponin T binds to tropomyosin at the head-to-tail junction.

(Images [a] and [b] courtesy of Linda Rost and David DeRosier, Brandeis University; [c] courtesy of George Phillips, Rice University.)

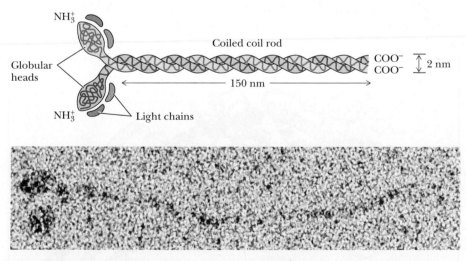

NH_3^+

Coiled coil rod

COO^-
COO^- 2 nm

Globular heads

150 nm

NH_3^+ Light chains

***Figure* 36.8**   An electron micrograph of a myosin molecule and a corresponding schematic drawing. The tail is a coiled coil of intertwined α-helices extending from the two globular heads. One of each of the myosin light chain proteins, LC1 and LC2, is bound to each of the globular heads.

(Electron micrograph courtesy of Henry Slayter, Harvard Medical School.)

Proteolysis of Myosin Produces Meromyosin Fragments

Treatment of myosin with trypsin was shown by Andrew Szent-Györgi in 1953 to produce two fragments, termed **heavy meromyosin (HMM)** and **light meromyosin (LMM);** Figure 36.9. HMM includes the myosin heads and a 55-kD segment of the tail. HMM binds light chains and possesses ATPase activity but does not form multimeric aggregates. LMM consists of long (85 nm) rods that spontaneously self-associate but do not bind light chains or hydrolyze ATP. HMM can be further treated with papain and other proteolytic enzymes to yield the globular heads, denoted **S1 fragments,** and long rods, called **S2**

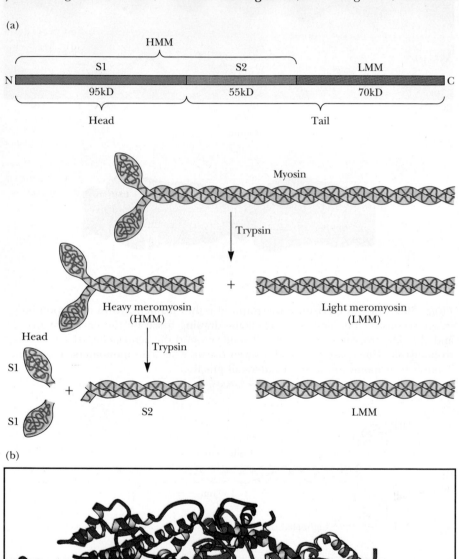

Figure **36.9**   Myosin dimers can be cleaved by trypsin treatment into light meromyosin and heavy meromyosin. Further proteolytic digestion cleaves the S1 heads from the S2 fragments. (a) A schematic of the primary sequence indicates the positions of cleavage and the lengths of the resulting fragments. (b) A ribbon diagram shows the structure of the S1 myosin head (green, red, and purple segments) and its associated essential (yellow) and regulatory (magenta) light chains.

(Image [b] courtesy of Ivan Rayment and Hazel M. Holden, University of Wisconsin, Madison.)

fragments. The S1 fragments retain the myosin ATPase activity, and they can also bind to actin in thin filaments.

Approximately 500 of the 820 amino acid residues of the myosin head are highly conserved between various species. One conserved region, located approximately at residues 170 to 214, constitutes part of the ATP-binding site. Whereas many ATP-binding proteins and enzymes employ a β-sheet–α-helix–β-sheet motif, this region of myosin forms a related α-β-α structure, beginning with an Arg at (approximately) residue 192. The β-sheet in this region of all myosins includes the amino acid sequence

Gly-Glu-Ser-Gly-Ala-Gly-Lys-Thr

The Gly-X-X-Gly-X-Gly found in this segment is found in many ATP- and nucleotide-binding enzymes. The Lys of this segment is thought to interact with the α-phosphate of bound ATP.

Repeating Structural Elements Are the Secret of Myosin's Coiled Coils

Myosin tails show less homology than the head regions, but several key features of the tail sequence are responsible for the unique α-helical coiled coils formed by myosin tails. *Several orders of repeating structure* are found in all myosin tails, including 7-residue, 28-residue, and 196-residue repeating units. Large stretches of the tail domain are comprised of 7-residue repeating segments. The first and fourth residues of these 7-residue units are generally small, hydrophobic amino acids, whereas the second, third, and sixth are likely to be charged residues. The consequence of this arrangement is shown in Figure 36.10. Seven residues form 2 turns of an α-helix, and, in the coiled coil structure of the myosin tails, the first and fourth residues face the interior contact region of the coiled coil. Residues 2, 3, and 6 of the 7-residue repeat face the periphery, where charged residues can interact with the water solvent. Groups of four 7-residue units with distinct patterns of alternating side-chain charge are also repeated in the myosin structure. These 28-residue repeats account for alternating regions of positive and negative charge on the surface of the myosin coiled coil. These alternating charged regions interact with the tails of adjacent myosin molecules to assist in stabilizing the thick filament.

At a still higher level of organization, groups of seven of these 28-residue units—a total of 196 residues—also form a repeating pattern, and this large-scale repeating motif contributes to the packing of the myosin molecules in the thick filament. The myosin molecules in thick filaments are offset (Figure 36.11) by approximately 14 nm, a distance that corresponds to 98 residues of

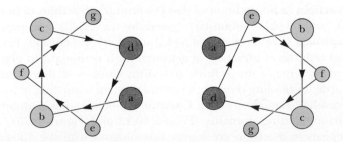

Figure **36.10**   An axial view of the two-stranded, α-helical coiled coil of a myosin tail. Hydrophobic residues 1 and 4 of the seven-residue repeat sequence align to form a hydrophobic core. Residues 2, 3, and 6 face the outer surface of the coiled coil and are typically ionic.

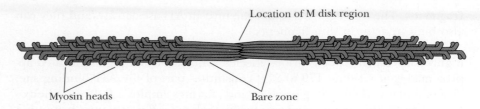

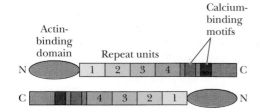

Figure 36.11 The packing of myosin molecules in a thick filament. Adjoining molecules are offset by approximately 14 nm, a distance corresponding to 98 residues of the coiled coil.

a coiled coil, or exactly half the length of the 196-residue repeat. Thus, several layers of repeating structure play specific roles in the formation and stabilization of the myosin coiled coil and the thick filament formed from them. In summary, the 7-residue repeats facilitate the arrangement of hydrophobic and charged residues on the interior faces and periphery of the coiled coil, respectively, and the 28-residue and 196-residue repeats produce regions of alternating positive and negative charges and assist in the packing of individual myosin molecules in the thick filament.

The Associated Proteins of Striated Muscle

In addition to the major proteins of striated muscle (myosin, actin, tropomyosin, and the troponins), numerous other proteins play important roles in the maintenance of muscle structure and the regulation of muscle contraction. Myosin and actin together account for 65% of the total muscle protein, and tropomyosin and the troponins each contribute an additional 5% (Table 36.1). The other regulatory and structural proteins thus comprise approximately 25% of the myofibrillar protein. The regulatory proteins can be classified as either **myosin-associated proteins** or as **actin-associated proteins.**

The myosin-associated proteins include three proteins found in the M disks. The M disks consist primarily of **M protein** (165 kD), **myomesin** (185 kD), and **creatine kinase** (a dimer of 42-kD subunits). Creatine kinase facilitates rapid regeneration of the ATP consumed during muscle contraction. The association of M protein, myomesin, and creatine kinase in the M disk maintains the structural integrity of the myosin filaments. Several other myosin-associated proteins have also been identified, including **C protein** (135 kD), **F protein** (121 kD), **H protein** (74 kD), and **I protein** (50 kD). The C protein is localized to several regularly spaced stripes in the A band. C protein inhibits myosin ATPase activity at low ionic strength but activates it at physiological ionic strength. The roles of F, H, and I proteins are not yet understood.

Actin-associated proteins (other than tropomyosin and the troponins) include **α-actinin** (a homodimer of 95-kD subunits), **β-actinin** (a heterodimer of 37-kD and 34-kD subunits), **γ-actinin** (a 35-kD monomer), and **paratropomyosin** (a homodimer of 34-kD subunits). α-Actinin is found in the Z lines and activates contraction of actomyosin. It is thought to play a role in attachment of actin to the Z lines. α-Actinin consists of three domains, an N-terminal, actin-binding domain; a central domain consisting of four repeats of a 122-residue sequence; and a C-terminal domain that contains two EF-hand, calcium-binding domains (Figure 36.12 and Figure 5.25). The four central repeats in α-actinin are highly homologous with the 106-residue repeat sequences of **spectrin,** the major structural protein of the red blood cell cytoskeleton. The repeating segments of both α-actinin and spectrin are thought to consist of bundles of four α-helices (Figure 36.13). β-Actinin acts

Figure 36.12 α-Actinin exists as a homodimer of antiparallel subunits, illustrated here in terms of their primary structure. The N-terminal, actin-binding domain and the C-terminal, EF-hand domains are separated by a central domain consisting of four repeats of a 122-residue sequence.

Table 36.1
Myofibrillar Structural Proteins of Rabbit Skeletal Muscle

Protein	Molecular Mass (kD)	Content (wt %)	Localization	Function
Contractile proteins				
Myosin	520	43	A band	Contracts with actin
Actin	42	22	I band	Contracts with myosin
Regulatory proteins				
Major				
Tropomyosin	33×2	5	I band	Binds to actin and locates troponin
Troponin	70	5	I band	Ca regulation
Troponin C	18			Ca binding
Troponin I	21			Inhibits actin-myosin interaction
Troponin T	31			Binds to tropomyosin
Minor				
M protein	165	2	M line	Binds to myosin
Myomesin	185	<1	M line	Binds to myosin
Creatine kinase	42	<1	M line	Binds to myosin
C protein	135	2	A band	Binds to myosin
F protein	121	<1	A band	Binds to myosin
H protein	74	<1	Near M line	Binds to myosin
I protein	50	<1	A band	Inhibits actin-myosin interaction
α-Actinin	95×2	2	Z line	Gelates actin filaments
β-Actinin	$37 + 34$	<1	Free end of actin filament	Caps actin filaments
γ-Actinin	35	<1	?	Inhibits actin polymerization
eu-Actinin	42	<1	Z line	Binds to actin
ABP (filamin)	240×2	<1	Z line	Gelates actin filaments
Paratropomyosin	34×2	<1	A–I junction	Inhibits actin-myosin interaction
Cytoskeletal proteins				
α-Connectin (titin 1)	2800	10	A–I	Links myosin filament to Z line
β-Connectin (titin 2)	2100			
Nebulin	800	5	N_2 line*	
Vinculin	130	<1	Under sarcolemma	
Desmin (skeletin)	53	<1	Periphery of Z line	Intermediate filament
Vimentin	55	<1	Periphery of Z line	Intermediate filament
Synemin	220	<1	Z line	
Z protein	50	<1	Z line	Forms lattice structure
Z-nin	400	<1	Z line	

Adapted from Ohtsuki, I., Maruyama, K., and Ebashi, S., 1986. Regulatory and cytoskeletal proteins of vertebrate skeletal muscle. *Advances in Protein Chemistry* **38**:1–67.

*A structure within the I band.

as an actin-capping protein, specifically binding to the end of an actin filament. β-Actinin may inhibit further elongation of actin polymers and interfilament interactions by capping the free ends of thin filaments. γ-Actinin also inhibits actin polymerization, but its location in thin filaments is not known with certainty. Paratropomyosin is similar to tropomyosin, but appears to be located only at the A band–I band junction.

Critical Developments in Biochemistry

The Molecular Defect in Duchenne Muscular Dystrophy Involves an Actin-Anchoring Protein

Discovery of a new actinin/spectrin-like protein has provided insights into the molecular basis for at least one form of muscular dystrophy. Duchenne muscular dystrophy is a degenerative and fatal disorder of muscle affecting approxi-mately 1 in 3500 boys. Victims of Du-chenne dystrophy show early abnormali-ties in walking and running. By the age of five, the victim cannot run and has difficulty standing, and by early adoles-cence, walking is difficult or impossible. The loss of muscle function progresses upward in the body, affecting next the arms and the diaphragm. Respiratory problems or infections usually result in death by the age of 30. Louis Kunkel and his co-workers identified the Du-chenne muscular dystrophy gene in 1986. This gene produces a protein called **dystrophin,** which is highly ho-mologous to α-actinin and spectrin. A defect in dystrophin is responsible for the muscle degeneration of Duchenne dystrophy.

Dystrophin is located on the cyto-plasmic face of the muscle plasma membrane, linked to the plasma mem-brane via an integral membrane glyco-protein. Dystrophin has a high molecu-

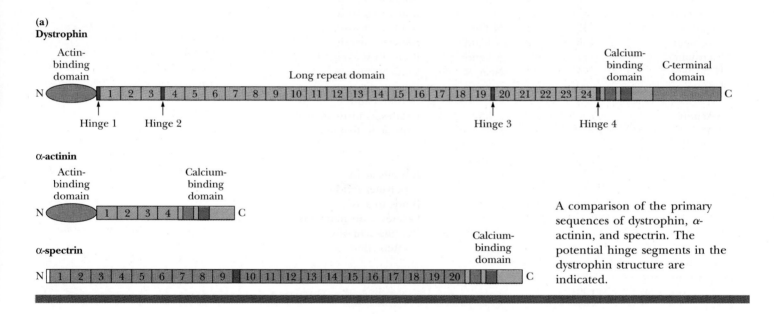

A comparison of the primary sequences of dystrophin, α-actinin, and spectrin. The potential hinge segments in the dystrophin structure are indicated.

Two cytoskeletal proteins, **connectin** (also known as **titin**) and **nebulin,** account for 15% of the total protein in the myofibril. Together these proteins form a flexible filamentous network that surrounds the myofibrils. Connectin is an **elastic protein** and can stretch under tension. Its discovery and charac-

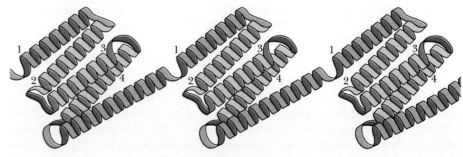

Figure 36.13 A schematic drawing of the four-helix cluster model for α-actinin and spectrin. Helix 1 is long and is postulated to lie at an angle with respect to the long axis of the repeated domain.

lar mass (427 kD), but constitutes less than 0.01% of the total muscle protein. It folds into four principal domains (figure, part a), including an N-terminal domain similar to the actin-binding domains of actinin and spectrin, a long repeat domain, a cysteine-rich domain, and a C-terminal domain that is unique to dystrophin. The repeat domain consists of 24 repeat units of approximately 109 residues each. "Spacer sequences" high in proline content, which do not

align with the repeat consensus sequence, occur at the beginning and end of the repeat domain. Spacer segments are found between repeat elements 3 and 4 and 19 and 20. The high proline content of the spacers suggests that they may represent hinge domains. The spacer/hinge segments are sensitive to proteolytic enzymes, indicating that they may represent more exposed regions of the polypeptide. Kunkel has proposed that

dystrophin may anchor cytoskeletal actin filaments to the plasma membrane via specific interactions with the glycoprotein complex (figure, part b).

(b)

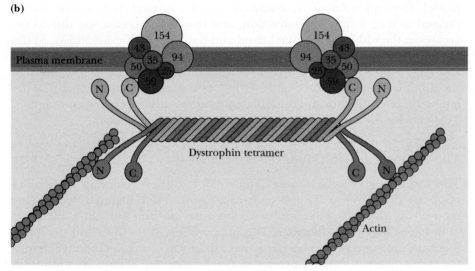

A model for the actin-dystrophin-glycoprotein complex in skeletal muscle. Dystrophin is postulated to form tetramers of antiparallel monomers that bind actin at their N-termini and a family of dystrophin-associated glycoproteins at their C-termini. This dystrophin-anchored complex may function to stabilize the sarcolemmal membrane during contraction-relaxation cycles, link the contractile force generated in the cell (fiber) with the extracellular environment, or maintain local organization of key proteins in the membrane. The dystrophin-associated membrane proteins range from 25 to 154 kD.

(Adapted from Ahn, A. H., and Kunkel, L. M., 1993, Nature Genetics 3, 283–291.)

terization ended a century-long debate over the possible existence of an elastic component in muscle. Connectin consists of two proteins: **α-connectin** and **β-connectin** (also called **titin 1** and **titin 2**).

The connectins are very large proteins: the mass of α-connectin is 2800 kD and that of β-connectin is 2100 kD. β-Connectin may be a proteolytic product of α-connectin. β-Connectin forms long, flexible, thin filaments. *A single β-connectin filament is approximately 1000 nm long in its relaxed state. Under tension, a single β-connectin filament can stretch to a length of over 3000 nm!* Connectin filaments in muscle originate at the periphery of the M band and extend along the myosin filaments all the way to the Z line (Figure 36.14). Their function appears to be to link the myosin filaments to the Z lines. When myofibrils are stretched beyond the overlap of the thick and thin filaments, connectin filaments passively generate tension. This effect is provided by a relatively small number of connectin molecules. The ratio of myosin to connectin filaments is approximately 24 to 1. With 300 myosin molecules per thick filament, only 6 or so connectin filaments are present in each half of a myosin thick filament.

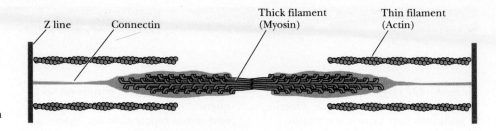

***Figure* 36.14**   A drawing of the arrangement of the elastic protein connectin in the skeletal muscle sarcomere. Connectin filaments originate at the periphery of the M band and extend along the myosin filaments to the Z lines. These connectin filaments produce the passive tension existing in myofibrils that have been stretched so that the thick and thin filaments no longer overlap and cannot interact.

(Adapted from Ohtsuki, I., Maruyama, K., and Ebashi, S., 1986. Advances in Protein Chemistry 38:1–67.)

36.3 The Mechanism of Muscle Contraction

When muscle fibers contract, the thick myosin filaments slide or walk along the thin actin filaments. The basic elements of the **sliding filament model** were first described in 1954 by two different research groups, Hugh Huxley and his colleague Jean Hanson, and the physiologist Andrew Huxley and his colleague Ralph Niedergerke. Several key discoveries paved the way for this model. Electron microscopic studies of muscle revealed that sarcomeres decreased in length during contraction, and that this decrease was due to decreases in the width of both the I band and the H zone (Figure 36.15). At the same time, the width of the A band (which is the length of the thick filaments) and the distance from the Z disks to the nearby H zone (that is, the length of the thin filaments) did not change. These observations made it clear that the lengths of both the thin and thick filaments were constant during contraction. This conclusion was consistent with a sliding filament model.

The Sliding Filament Model

The shortening of a sarcomere (Figure 36.15) involves sliding motions in opposing directions at the two ends of a myosin thick filament. Net sliding motions in a specific direction occur because the thin and thick filaments both have **directional character.** The organization of the thin and thick filaments in the sarcomere takes particular advantage of this directional character. Actin filaments always extend outward from the Z lines in a uniform

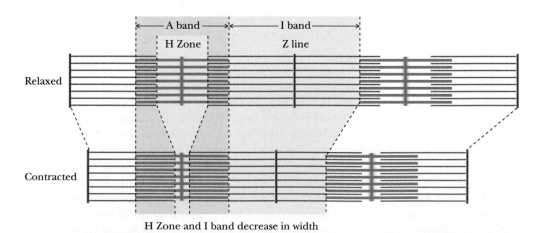

***Figure* 36.15**   The sliding filament model of skeletal muscle contraction. The decrease in sarcomere length is due to decreases in the width of the I band and H zone, with no change in the width of the A band. These observations mean that the lengths of both the thick and thin filaments do not change during contraction. Rather, the thick and thin filaments slide along one another.

A Deeper Look

Viscous Solutions Reflect Long-Range Molecular Interactions

High viscosity in an aqueous solution is a sign of long-range molecular interactions, defined as interactions that extend through and connect many molecules. Concentrated sugar solutions (molasses, for example) are viscous because of the extensive hydrogen-bonding networks established by the multiple hydroxyl groups of sugar molecules. Solutions of DNA are highly viscous because isolated DNA fibers are extremely long (often in the millimeter size range) and highly hydrated. As Szent-Györgyi discovered, the extensive aggregates formed by myosin and actin also produce highly viscous solutions.

---

manner. Thus, between any two Z lines, the two sets of actin filaments point in opposing directions. The myosin thick filaments, on the other hand, also assemble in a directional manner. The polarity of myosin thick filaments reverses at the M disk. The nature of this reversal is not well understood, but presumably involves structural constraints provided by proteins in the M disk, such as the M protein and myomesin described above. The reversal of polarity at the M disk means that actin filaments on either side of the M disk are pulled toward the M disk during contraction by the sliding of the myosin heads, causing net shortening of the sarcomere.

Albert Szent-Györgyi's Discovery of the Effects of Actin on Myosin

The molecular events of contraction are powered by the ATPase activity of myosin. Much of our present understanding of this reaction and its dependence on actin can be traced to several key discoveries by Albert Szent-Györgyi at the University of Szeged in Hungary in the early 1940s. Szent-Györgyi showed that solution viscosity is dramatically increased when solutions of myosin and actin are mixed. Increased viscosity is a manifestation of the formation of an **actomyosin complex.**

Szent-Györgyi further showed that the viscosity of an actomyosin solution was lowered by the addition of ATP, indicating that ATP decreases myosin's affinity for actin. Kinetic studies demonstrated that myosin ATPase activity was increased substantially by actin. (For this reason, Szent-Györgyi gave the name **actin** to the thin filament protein.) The ATPase turnover number of pure myosin is 0.05/sec. In the presence of actin, however, the turnover number increases to about 10/sec, a number more like that of intact muscle fibers.

The specific effect of actin on myosin ATPase becomes apparent if the *product release* steps of the reaction are carefully compared. In the absence of actin, the addition of ATP to myosin produces a rapid release of H^+, one of the products of the ATPase reaction:

$$ATP^{4-} + H_2O \rightarrow ADP^{3-} + P_i^{2-} + H^+$$

However, release of ADP and P_i from myosin is much slower. Actin activates myosin ATPase activity by stimulating the release of P_i and then ADP. Product release is followed by the binding of a new ATP to the actomyosin complex, which causes actomyosin to dissociate into free actin and myosin. The cycle of ATP hydrolysis then repeats, as shown in Figure 36.16. The crucial point of this model is that *ATP hydrolysis and the association and dissociation of actin and myosin are coupled.* It is this coupling that enables ATP hydrolysis to power muscle contraction.

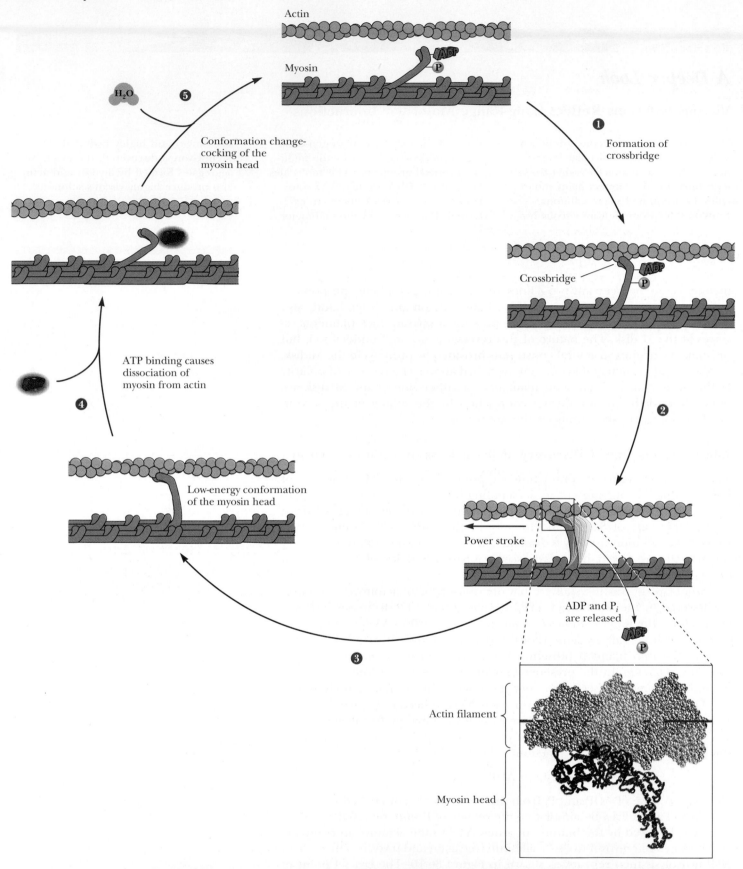

Actin

Myosin

H₂O

❺ Conformation change-cocking of the myosin head

❶ Formation of crossbridge

Crossbridge

ATP binding causes dissociation of myosin from actin

❹

Low-energy conformation of the myosin head

Power stroke

❷

❸

ADP and P$_i$ are released

Actin filament

Myosin head

Figure 36.16 The mechanism of skeletal muscle contraction. The free energy of ATP hydrolysis drives a conformation change in the myosin head, resulting in net movement of the myosin heads along the actin filament. *Inset:* A ribbon and space-filling representation of the actin-myosin interaction.

(Image courtesy of Ivan Rayment and Hazel M. Holden, University of Wisconsin, Madison.)

The Coupling Mechanism: ATP Hydrolysis Drives Conformation Changes in the Myosin Heads

The only remaining piece of the puzzle is this: How does the close coupling of actin-myosin binding and ATP hydrolysis result in the shortening of myofibrils? Put another way, how are the model for ATPase hydrolysis and the sliding filament model related? The answer to this puzzle is shown in Figure 36.16. The free energy of ATP hydrolysis is translated into a conformation change in the myosin head, so that dissociation of myosin and actin, hydrolysis of ATP, and rebinding of myosin and actin occurs with stepwise movement of the myosin S1 head along the actin filament. The conformation change in the myosin head is driven by the hydrolysis of ATP.

As shown in the cycle in Figure 36.16, the myosin heads—with the hydrolysis products ADP and P_i bound—are mainly dissociated from the actin filaments in resting muscle. When the signal to contract is presented (see following discussion), the myosin heads move out from the thick filaments to bind to actin on the thin filaments (Step 1). Binding to actin stimulates the release of phosphate and this is followed by the crucial conformational change by the S1 myosin heads—the so-called **power stroke**—and ADP dissociation. In this step (Step 2), the thick filaments *move* along the thin filaments as the myosin heads relax to a lower energy conformation. In the power stroke, the myosin heads tilt by approximately 45° and the conformational energy of the myosin heads is lowered by about 29 kJ/mol. This moves the thick filament approximately 10 nm along the thin filament (Step 3). Subsequent binding (Step 4) and hydrolysis (Step 5) of ATP cause dissociation of the heads from the thin filaments and also cause the myosin heads to shift back to their high energy conformation with the heads' long axis nearly perpendicular to the long axis of the thick filaments. The heads may then begin another cycle by binding to actin filaments. This cycle is repeated at rates up to 5/sec in a typical skeletal muscle contraction. The conformational changes occurring in this cycle are the secret of the energy coupling that allows ATP binding and hydrolysis to drive muscle contraction.

36.4 Control of the Contraction-Relaxation Cycle by Calcium Channels and Pumps

The trigger for all muscle contraction is an increase in Ca^{2+} concentration in the vicinity of the muscle fibers of skeletal muscle or the myocytes of cardiac and smooth muscle. In all these cases, this increase in Ca^{2+} is due to the flow of Ca^{2+} through **calcium channels** (Figure 36.17). A muscle contraction ends when the Ca^{2+} concentration is reduced by specific calcium pumps (such as the SR Ca^{2+}-ATPase, Chapter 35). The sarcoplasmic reticulum, t-tubule, and sarcolemmal membranes all contain Ca^{2+} channels. As we shall see, the Ca^{2+} channels of the SR function together with the t-tubules in a remarkable coupled process.

Ca^{2+} release in skeletal and heart muscle has been characterized through the use of specific antagonist molecules which block Ca^{2+} channel activity. The **dihydropyridine (DHP) receptors** of t-tubules, for example, are blocked by **dihydropyridine** derivatives, such as **nifedipine** (Figure 36.18). The purified DHP receptor of heart muscle can be incorporated into liposomes, whereupon it shows calcium channel activity. The channel displays voltage-dependent gating and is selective for divalent cations over monovalent cations. *Thus, the heart muscle DHP receptor is a voltage-dependent Ca^{2+} channel.* Other evidence suggests that the skeletal muscle DHP receptor is a voltage-sensing protein; it presumably undergoes voltage-dependent conformation changes.

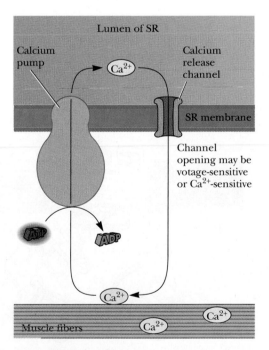

Figure 36.17 Ca^{2+} is the trigger signal for muscle contraction. Release of Ca^{2+} through voltage- or Ca^{2+}-sensitive channels activates contraction. Ca^{2+} pumps induce relaxation by reducing the concentration of Ca^{2+} available to the muscle fibers.

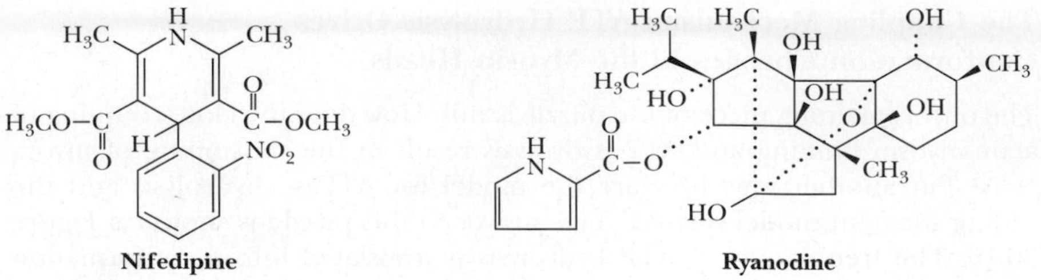

Nifedipine

Ryanodine

Figure 36.18 The structures of nifedipine and ryanodine. Nifedipine binds with high affinity to the Ca^{2+}-release channels of t-tubules. Ryanodine binds with high affinity to the Ca^{2+} channels of SR terminal cisternae.

The DHP receptor from t-tubules consists of five different polypeptides, designated α_1 (150 to 173 kD), α_2 (120 to 150 kD), β (50 to 65 kD), γ (30 to 35 kD), and δ (22 to 27 kD). The α_2- and δ-subunits are linked by a disulfide bond. The α_1, α_2-δ, β, and γ stoichiometry is $1:1:1:1$. The α_2-subunit is glycosylated, but α_1 is not. α_1 is homologous with the α-subunit of the voltage-sensitive sodium channel (Chapter 38). The sequence of α_1 contains four internal sequence repeats, each containing six transmembrane helices, one of which is positively charged and which is believed to be a voltage sensor (Figure 36.19). The loop between helices 5 and 6 contributes to the pore. These six segments share many similarities with the corresponding segments of the sodium channel. The α_1-subunit of the DHP receptor in heart muscle is implicated in channel formation and voltage-dependent gating.

The Ca^{2+}-release channel from the terminal cisternae of sarcoplasmic reticulum has been identified by virtue of its high affinity for **ryanodine,** a toxic alkaloid (Figure 36.18). The purified receptor consists of oligomers, containing four or more subunits of a single large polypeptide (565 kD).

DHP-sensitive calcium channel

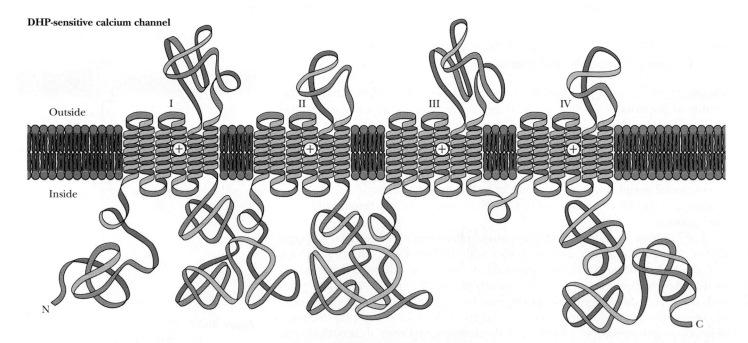

Outside

Inside

Figure 36.19 The α_1-subunit of the t-tubule Ca^{2+} channel/DHP receptor contains six peptide segments that may associate to form the Ca^{2+} channel. This Ca^{2+} channel polypeptide is homologous with the voltage-sensitive Na^+ channel of neuronal tissue.

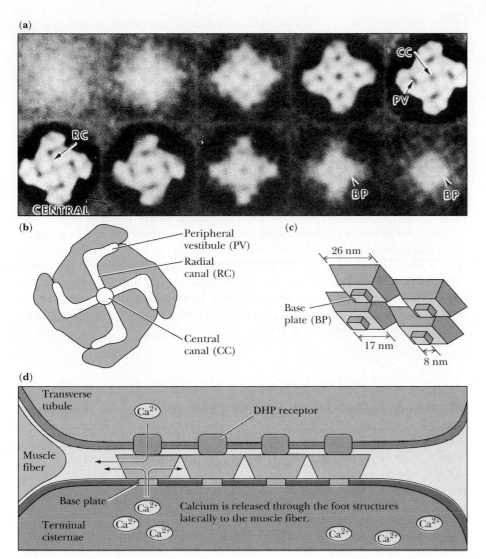

(a)

(b)

Peripheral vestibule (PV)

Radial canal (RC)

Central canal (CC)

(c)

26 nm

Base plate (BP)

17 nm

8 nm

(d)

Transverse tubule

Ca^{2+}

DHP receptor

Muscle fiber

Base plate

Ca^{2+}

Calcium is released through the foot structures laterally to the muscle fiber.

Ca^{2+}

Ca^{2+}

Ca^{2+}

Ca^{2+}

Ca^{2+}

Ca^{2+}

Terminal cisternae

Figure **36.20** (a) Electron micrograph images of foot structures of terminal cisternae. (b,c) Foot structures appear as trapezoids and diamonds on the surface of the membrane. The central canal (CC), radial canals (RC), and peripheral vestibules (PV) are indicated. (d) The relationship between the foot structures, t-tubule, terminal cisternae, and muscle fiber.

Electron microscopy reveals that the purified ryanodine receptor (Figure 36.20) is in fact the **foot structure** observed in native muscle tissue. Image reconstructions of electron microscopic data reveal that the receptor is a square structure with fourfold symmetry. Reconstructed images show that the structure contains a central pore with four radially extending canals (Figure 36.21). These radial canals each extend to openings in the periphery of the structure and are therefore contiguous with the myoplasm.

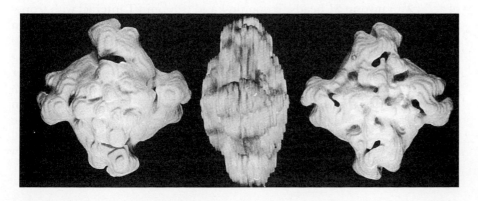

Figure **36.21** Image reconstructions of the junctional channel complex of a foot structure.

So how do the foot structures effect the release of Ca^{2+} from the terminal cisternae of the SR? The feet that join the t-tubules and the terminal cisternae of the SR are approximately 16 nm thick. The feet apparently function by first sensing either a voltage-dependent conformation change (skeletal muscle) or the transport of Ca^{2+} across the voltage-sensitive Ca^{2+} channel (heart muscle) of the t-tubule and then facilitating the release of large amounts of Ca^{2+} through the foot structure from the SR. The reconstructed image (Figure 36.21) for the foot structure suggests a possible pathway for Ca^{2+} transport from the lumen of the SR to the myoplasm via the ryanodine receptor. A Ca^{2+}- or voltage-dependent conformation change may serve to gate open the central canal of the foot structure. On entering the central canal, calcium ions move outward through the radial canals to the outer vestibule regions and into the myoplasm adjacent to the triad junctions, where binding to the muscle fibers induces contraction. Many of the details of these processes are presently unknown, and considerable additional work will be required to completely elucidate the mechanisms of the voltage- and Ca^{2+}-dependent Ca^{2+} release into the myoplasm that triggers muscle contraction. By contrast, the details of the contraction process are much better understood, especially in skeletal muscle. The following sections will describe the molecular events that enables muscle contraction to occur.

36.5 Regulation of Contraction by Ca^{2+}

Early in this chapter, the importance of Ca^{2+} ion as the triggering signal for muscle contraction was described. Ca^{2+} is the intermediary signal that allows striated muscle to respond to motor nerve impulses (Figure 36.17). The importance of Ca^{2+} as a contraction signal was understood in the 1940s, but it remained for Setsuro Ebashi, a pioneer of muscle research, to show in the early 1960s that the Ca^{2+} signal is correctly interpreted by muscle only when tropomyosin and the troponins are present. Specifically, actomyosin prepared from pure preparations of actin and myosin (thus containing no tropomyosin and troponins) was observed to contract when ATP was added, even in the absence of Ca^{2+}. However, actomyosin prepared directly from whole muscle would contract in the presence of ATP only when Ca^{2+} was added. Clearly the muscle extracts contained a factor that conferred normal Ca^{2+} sensitivity to actomyosin. The factor turned out to be the tropomyosin-troponin complex.

Actin thin filaments consist of actin, tropomyosin, and the troponins in a 7:1:1 ratio (Figure 36.7). Each tropomyosin molecule spans seven actin molecules, lying along the thin filament groove, between pairs of actin monomers. As shown in a cross-section view in Figure 36.22, in the absence of Ca^{2+}, troponin I is thought to interact directly with actin to prevent the interaction of actin with myosin S1 heads. Troponin I and troponin T interact with tropomyosin to keep tropomyosin away from the groove between adjacent actin monomers. However, the binding of Ca^{2+} ions to troponin C appears to increase the binding of troponin C to troponin I, simultaneously decreasing the interaction of troponin I with actin. As a result, tropomyosin slides deeper into the actin-thin filament groove, exposing myosin-binding sites on actin, and initiating the muscle contraction cycle (Figure 36.16). Since the troponin complexes can interact only with every seventh actin in the thin filament, the conformational changes that expose myosin-binding sites on actin may well be cooperative. Binding of an S1 head to an actin may displace tropomyosin and the troponin complex from myosin-binding sites on adjacent actin subunits.

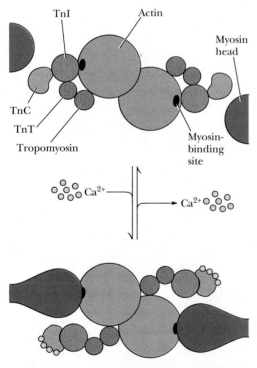

Figure 36.22 A drawing of the thick and thin filaments of skeletal muscle in cross-section showing the changes that are postulated to occur when Ca^{2+} binds to troponin C.

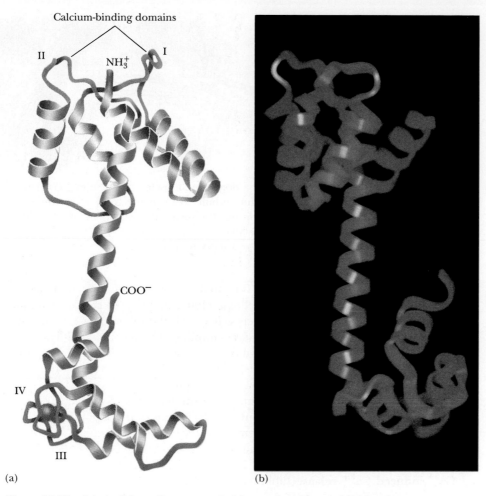

Calcium-binding domains

II   I

NH$_3^+$

COO$^-$

IV

III

(a)

(b)

***Figure* 36.23**   (a) A ribbon diagram and (b) a molecular graphic showing two slightly different views of the structure of troponin C. Note the long α-helical domain connecting the N-terminal and C-terminal lobes of the molecule.

The Interaction of Ca^{2+} with Troponin C

There are four Ca^{2+}-binding sites on troponin C—two high-affinity sites on the carboxy-terminal end of the molecule, labeled III and IV in Figure 36.23, and two low-affinity sites on the amino-terminal end, labeled I and II. Ca^{2+} binding to sites III and IV is sufficiently strong ($K_D = 0.1~\mu M$) that these sites are presumed to be filled under resting conditions. Sites I and II, however, where the K_D is approximately 10 μM, are empty in resting muscle. The rise of Ca^{2+} levels when contraction is signaled leads to the filling of sites I and II, causing a conformation change in the amino-terminal domain of TnC. This conformational change apparently facilitates a more intimate binding of TnI to TnC that involves the C helix, and also possibly the E helix of TnC. The increased interaction between TnI and TnC results in a decreased interaction between TnI and actin.

36.6 The Structure of Cardiac and Smooth Muscle

The structure of heart myocytes is different from that of skeletal muscle fibers. Heart myocytes are approximately 50 to 100 μm long and 10 to 20 μm in diameter. The t-tubules found in heart tissue have a fivefold larger diameter than those of skeletal muscle. The number of t-tubules found in cardiac mus-

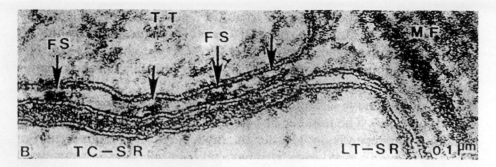

Figure 36.24   Electron micrograph of a dog heart muscle. The terminal cisterna of the SR (TC-SR) is associated with the t-tubule (TT) by means of foot structures (FS), forming a dyad junction. MF indicates the location of myofilaments. LT-SR signifies the longitudinal tubule of the SR.
(From: Fleischer, S., and Inui, M., 1989. Annual Review of Biophysics and Biophysical Chemistry 18:333–364.)

cle differs from species to species. Terminal cisternae of mammalian cardiac muscle can associate with other cellular elements to form **dyads** as well as triads. The association of terminal cisternae with the sarcolemma membrane in a dyad structure is called a **peripheral coupling.** The terminal cisternae may also form dyad structures with t-tubules that are called **internal couplings** (Figure 36.24). As with skeletal muscle, foot structures form the connection between the terminal cisternae and t-tubule membranes.

In higher animals, large percentages of the terminal cisternae of cardiac muscle are not associated with t-tubules at all. For SR of this type, Ca^{2+} release must occur by a different mechanism from that found in skeletal muscle. In this case, it appears that Ca^{2+} leaking through sarcolemma Ca^{2+} channels can trigger the release of even more Ca^{2+} from the SR. This latter process is called **Ca^{2+}-induced Ca^{2+} release** (abbreviated **CICR**).

The Structure of Smooth Muscle Myocytes

The myocytes of smooth muscle are approximately 100 to 500 μm in length and only 2 to 6 μm in diameter. Smooth muscle contains very few t-tubules and much less SR than skeletal muscle. The Ca^{2+} that stimulates contraction in smooth muscle cells is predominantly extracellular in origin. This Ca^{2+} enters the cell through Ca^{2+} channels in the sarcolemmal membrane that can be opened by electrical stimulation, or by the binding of hormones or drugs. The contraction response time of smooth muscle cells is very slow compared with that of skeletal and cardiac muscle.

36.7 The Mechanism of Smooth Muscle Contraction

Vertebrate organisms employ smooth muscle myocytes for long, slow, and involuntary contractions in various organs, including large blood vessels; intestinal walls; and, in the female, the uterus. Smooth muscle contains no troponin complex; thin filaments consist only of actin and tropomyosin. Despite the absence of troponins, smooth muscle contraction *is* dependent on Ca^{2+}, which activates **myosin light chain kinase (MLCK),** an enzyme that phosphorylates LC2, the regulatory light chain of myosin. Contraction of smooth muscle is initiated by phosphorylation of LC2, and dephosphorylation causes relaxation of smooth muscle tissue.

The mechanism of this contraction process is shown in Figure 36.25. Smooth muscle myocytes have a resting $[Ca^{2+}]$ of approximately 0.1 μM. Electrical stimulation (by the autonomic or involuntary nervous system) opens Ca^{2+} channels in the sarcolemma membrane, allowing $[Ca^{2+}]$ to rise to about 10 μM, a concentration at which Ca^{2+} binds readily to a protein called **calmodulin** (see Chapter 37). Binding of the Ca^{2+}-calmodulin complex to MLCK activates the kinase reaction, phosphorylating LC2 and stimulating smooth muscle contraction. Export of Ca^{2+} by the plasma membrane Ca^{2+}-ATPase returns Ca^{2+} to its resting level, deactivating MLCK. Smooth muscle relaxation then occurs through the action of **myosin light chain phosphatase,** which dephosphorylates LC2. This reaction is relatively slow, and smooth muscle contractions are typically more sustained and dissipate more slowly than those of striated muscle.

Smooth muscle contractions are subject to the actions of hormones and related agents. As shown in Figure 36.25, binding of the hormone **epinephrine** to smooth muscle receptors activates an intracellular **adenylyl cyclase** reaction that produces cyclic AMP (cAMP). The cAMP serves to activate a protein kinase that phosphorylates the myosin light chain kinase. The phosphorylated MLCK has a lower affinity for the Ca^{2+}-calmodulin complex and thus is physiologically inactive. The reversal of this inactivation occurs via a specific **myosin light chain kinase phosphatase.**

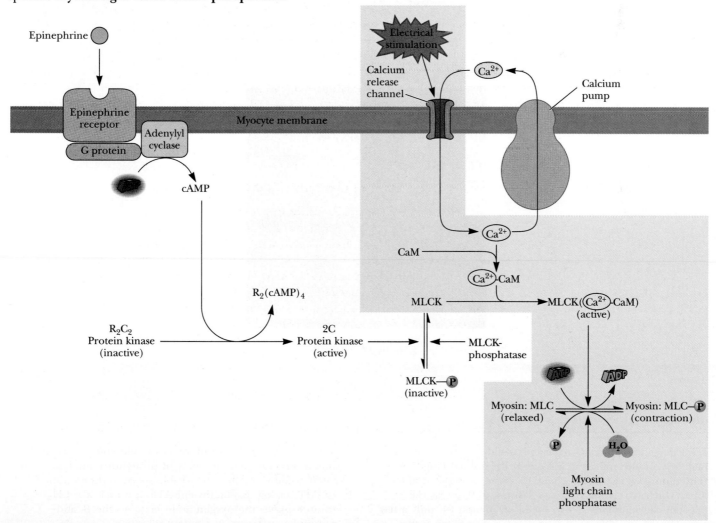

Figure **36.25** A model for the control of contraction in smooth muscle. (Calmodulin is abbreviated as CaM.)

A Deeper Look

Smooth Muscle Effectors Are Useful Drugs

The action of epinephrine and related agents forms the basis of therapeutic control of smooth muscle contraction. Breathing disorders, including asthma and various allergies, can result from excessive contraction of bronchial smooth muscle tissue. Treatment with epinephrine, whether by tablets or aerosol inhalation, inhibits MLCK and relaxes bronchial muscle tissue. More specific **bronchodilators,** such as **albuterol** (see figure), act more selectively on the lungs and avoid the undesirable side effects of epinephrine on the heart. Albuterol is also used to prevent premature labor in pregnant women, owing to its relaxing effect on uterine smooth muscle. Conversely, **oxytocin,** known also as **pitocin,** stimulates contraction of uterine smooth muscle. This natural secretion of the pituitary gland is often administered to induce labor.

Albuterol

$$H_3\overset{+}{N} - Gly - Leu - Pro - Cys - Asn - Gln - Ile - Tyr - Cys - COO^-$$
$$S\!-\!S$$

Oxytocin (Pitocin)

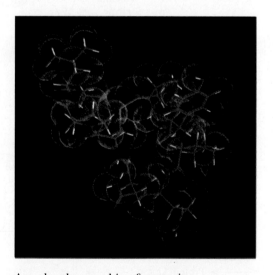

A molecular graphic of oxytocin.

Problems

1. The cheetah is generally regarded as nature's fastest mammal, but another amazing athlete in the animal kingdom (and almost as fast as the cheetah) is the pronghorn antelope, which roams the plains of Wyoming. Whereas the cheetah can maintain its top speed of 70 mph for only a few seconds, the pronghorn antelope can run at 60 mph for about an hour! (It is thought to have evolved to do so in order to elude now-extinct ancestral cheetahs that lived in North America.) What differences would you expect in the muscle structure and anatomy of pronghorn antelopes that could account for the remarkable speed and endurance?

2. An ATP analog, β,γ-methylene-ATP, in which a $-CH_2-$ group replaces the oxygen atom between the β- and γ-phosphorus atoms, is a potent inhibitor of muscle contraction. At which step in the contraction cycle would you expect β,γ-methylene-ATP to block contraction?

3. ATP stores in muscle are augmented or supplemented by stores of phosphocreatine. During periods of contraction, phosphocreatine is hydrolyzed to drive the synthesis of needed ATP in the creatine kinase reaction:

$$\text{phosphocreatine} + \text{ADP} \rightleftharpoons \text{creatine} + \text{ATP}$$

Muscle cells contain two different isozymes of creatine kinase, one in the mitochondria and one in the sarcoplasm. Explain.

4. *Rigor* is a muscle condition in which muscle fibers, depleted of ATP and phosphocreatine, develop a state of extreme rigidity and cannot be easily extended. (In death, this state is called *rigor mortis,* the rigor of death.) From what you have learned about muscle contraction, explain the state of rigor in molecular terms.

5. Skeletal muscle can generate approximately 3 to 4 kg of tension or force per square centimeter of cross-sectional area. This number is roughly the same for all mammals. Since many human muscles have large cross-sectional areas, the force that these muscles can (and must) generate is prodigious. The gluteus maximus (on which you are probably sitting as you read this) can generate a tension of 1200 kg! Estimate the cross-sectional area of all of the muscles in your body and the total force that your skeletal muscles could generate if they all contracted at once.

Further Reading

Ahn, A. H., and Kunkel, L. M., 1993. The structural and functional diversity of dystrophin. *Nature Genetics* **3**:283–291.

Amos, L., 1985. Structure of muscle filaments studied by electron microscopy. *Annual Review of Biophysics and Biophysical Chemistry* **14**:291–313.

Bagshaw, C., 1982. *Muscle Contraction.* London: Chapman and Hall.

Blanchard, A., Ohanian, V., and Critchley, D., 1989. The structure and function of α-actinin. *Journal of Muscle Research and Cell Motility* **10**:280–289.

Cooke, R., 1986. The mechanism of muscle contraction. *CRC Critical Reviews in Biochemistry* **21**:53–118.

Davison, M., and Critchley, D., 1988. α-Actinin and the DMD protein contain spectrin-like repeats. *Cell* **52**:159–160.

Davison, M., et al., 1989. Structural analysis of homologous repeated domains in α-actinin and spectrin. *International Journal of Biological Macromolecules* **11**:81–90.

Eisenberg, E., and Hill, T., 1985. Muscle contraction and free energy transduction in biological systems. *Science* **227**:999–1006.

Finer, J. T., Simmons, R. M., and Spudich, J. A., 1994. Single myosin molecule mechanics: Piconewton forces and nanometer steps. *Nature* **368**:113–119.

Fleischer, S., and Inui, M., 1989. Biochemistry and biophysics of excitation-contraction coupling. *Annual Review of Biophysics and Biophysical Chemistry* **18**:333–364.

Harrington, W., and Rodgers, M., 1984. Myosin. *Annual Review of Biochemistry* **53**:35–73.

Hibbard, M., and Trentham, D., 1986. Relationship between chemical and mechanical events during muscular contraction. *Annual Review of Biophysics and Biophysical Chemistry* **15**:119–161.

Kabsch, W., et al., 1990. Atomic structure of the actin:DNase I complex. *Nature* **347**:37–43.

Koenig, M., and Kunkel, L., 1990. Detailed analysis of the repeat domain of dystrophin reveals four potential hinge segments that may confer flexibility. *Journal of Biological Chemistry* **265**:4560–4566.

Korn, E., and Hammer, J., 1988. Myosins of non-muscle cells. *Annual Review of Biophysics and Biophysical Chemistry* **17**:23–45.

McLachlan, A., 1984. Structural implications of the myosin amino acid sequence. *Annual Review of Biophysics and Bioengineering* **13**:167–189.

Ohtsuki, I., Maruyama, K., and Ebashi, S., 1986. Regulatory and cytoskeletal proteins of vertebrate skeletal muscle. *Advances in Protein Chemistry* **38**:1–67.

Phillips, G., Fillers, J., and Cohen, C., 1986. Tropomyosin crystal structure and muscle regulation. *Journal of Molecular Biology* **192**:111–131.

Rayment, I., and Holden, H., 1994. The three-dimensional structure of a molecular motor. *Trends in Biochemical Sciences* **19**:129–134.

Saito, A., et al., 1988. Ultrastructure of the calcium release channel of sarcoplasmic reticulum. *Journal of Cell Biology* **107**:211–219.

Squire, J., 1986. *Muscle: Design, Diversity and Disease.* Menlo Park, CA: Benjamin/Cummings.

Squire, J., 1981. *The Structural Basis of Muscle Contraction.* New York: Plenum Press.

Tanabe, T., et al., 1987. Primary structure of the receptor for calcium channel blockers from skeletal muscle. *Nature* **328**:313–318.

Thomas, D., 1987. Spectroscopic probes of muscle crossbridge rotation. *Annual Review of Physiology* **49**:691–709.

Wagenknecht, T., et al., 1989. Three-dimensional architecture of the calcium channel/foot structure of sarcoplasmic reticulum. *Nature* **338**:167–170.

Warrick, H., and Spudich, J., 1987. Myosin: Structure and function in cell motility. *Annual Review of Cell Biology* **3**:379–422.

Zot, A., and Potter, J., 1987. Structural aspects of troponin-tropomyosin regulation of skeletal muscle contraction. *Annual Review of Biophysics and Biophysical Chemistry* **16**:535–559.

*"Ships that pass in the night, and
speak each other in passing,
Only a signal shown and a distant
voice in the darkness."*

"The Theologian's Tale," Henry Wadsworth
Longfellow (1807–1882)

Chapter 37

The Molecular Basis
of Hormone Action

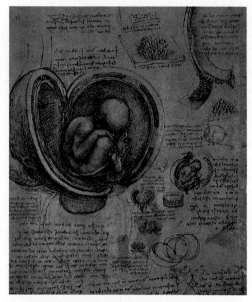

Drawing of a human fetus *in utero,* by Leonardo
da Vinci. Human sexuality and embryonic
development represent two hormonally regulated
processes of universal interest.

Outline

 igher life forms must have mechanisms for controlling and coordinating the many and diverse processes that occur in different parts of the organism. **Hormones** are chemical signals that provide this coordination. Hormones are secreted by certain cells, usually located in glands, and travel, either by simple diffusion or circulation in the bloodstream, to specific target cells. As we shall see, some hormones bind to specialized receptors on the plasma membrane and induce responses within the cell without themselves entering the target cell. Other hormones actually enter the target cell and interact with specific receptors there. By these mechanisms, hormones regulate the metabolic processes of various organs and tissues; facilitate and control growth, differentiation, and reproductive activities; and help the organism cope with changing conditions and stresses in its environment.

37.1 Classes of Hormones

Many different chemical species act as hormones. **Steroid hormones,** all derived from cholesterol, regulate metabolism, salt and water balances, inflammatory processes, and sexual function. Several hormones are **amino acid derivatives.** Among these are *epinephrine* and *norepinephrine,* which regulate smooth muscle contraction and relaxation, blood pressure, cardiac rate, and the processes of lipolysis and glycogenolysis, and the *thyroid hormones,* which stimulate metabolism. **Peptide hormones** are a large and still expanding group of hormones that appear to regulate processes in all body tissues, including the release of yet other hormones.

Hormones and other signal molecules in biological systems bind with very high affinities to their receptors, displaying K_D values in the range of 10^{-12} to 10^{-6} M. The concentrations of hormones are maintained at levels equivalent to or slightly above these K_D values. Once hormonal effects have been induced, the hormone is usually rapidly metabolized. The low concentrations and short lifetimes of secreted hormones have typically made identification of hormones and elucidation of their mechanisms of action extremely difficult. Modern biochemical methods, including molecular biological methods for cloning and expression of the genes encoding hormone receptors and radioimmunoassay techniques for detection and quantitation of hormones, have stimulated rapid advances in this field in recent years.

37.2 Signal-Transducing Receptors Transmit the Hormonal Message

Very often in life, *the message is more important than the messenger,* and this is certainly true for hormones. The structure and chemical properties of a hormone are only important for specific binding of the hormone to its appropriate receptor. Of much greater interest and importance, however, is the metabolic information carried by the hormonal signal. The information implicit in the hormonal signal is interpreted by the cell, and an intricate pattern of cellular response ensues.

Steroid hormones may either bind to plasma membrane receptors or exert their effects within target cells, entering the cell and migrating to their sites of action via specific cytoplasmic receptor proteins (Figure 37.1). The nonsteroid hormones, which act by binding to outward-facing plasma membrane receptors, activate various **signal transduction pathways** that mobilize various **second messengers**—cyclic nucleotides, Ca^{2+} ions, and other substances that activate or inhibit enzymes or cascades of enzymes in very specific ways. These hormonally activated processes are the focus of this chapter.

All receptors that mediate transmembrane signaling processes fit into one of three **receptor superfamilies:**

1. The **7-transmembrane segment (7-TMS) receptors** are integral membrane proteins with seven transmembrane (helical) segments, an extracellular recognition site for ligands, and an intracellular recognition site for a **GTP-binding protein** (see following discussion).

2. The **single-transmembrane segment catalytic receptors** are proteins with only a single transmembrane segment and substantial globular domains on both the extracellular and intracellular faces of the membrane. The extracellular domain is the ligand recognition site and the intracellular *catalytic* domain is either a **tyrosine kinase** or a **guanylyl cyclase.**

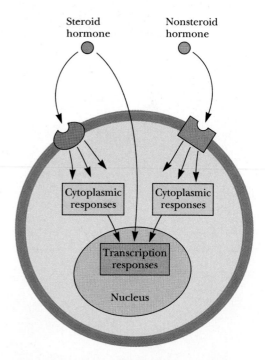

***Figure* 37.1** Nonsteroid hormones bind exclusively to plasma membrane receptors, which mediate cellular responses to the hormone. Steroid hormones exert their effects either by binding to plasma membrane receptors or by diffusing to the nucleus, where they modulate transcriptional events.

3. **Oligomeric ion channels** consist of associations of protein subunits, each of which contains several transmembrane segments. These oligomeric structures are **ligand-gated ion channels.** Binding of the specific ligand typically opens the ion channel. As discussed in Chapter 38, the ligands for these ion channels are *neurotransmitters.*

37.3 Intracellular Second Messengers

Cyclic AMP and the Second Messenger Model

Epinephrine and glucagon activate glycogen breakdown in the liver (Chapter 21), but the mechanism of this activation was a mystery until Earl Sutherland and his colleagues showed that the **glycogen phosphorylase** reaction (Figure 37.2), the initial step in glycogen breakdown, was stimulated by epinephrine and glucagon. Activation of phosphorylase was eventually shown to be due to an ATP-dependent phosphorylation of the enzyme. Sutherland also showed that a phosphatase from the liver cells inactivated phosphorylase, eliminat-

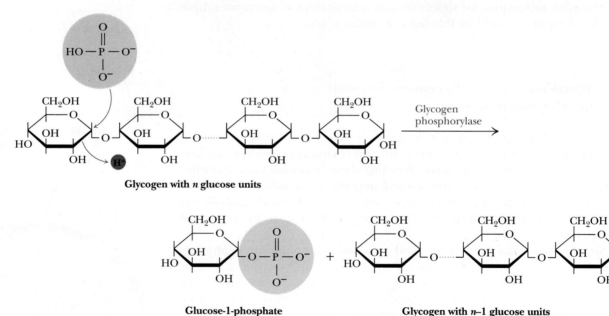

Figure 37.2   Glycogen breakdown in the liver.

ing the activation due to phosphorylation. A significant breakthrough was achieved with Sutherland's demonstration that hormones activated phosphorylase only in the presence of plasma membrane fragments. He hypothesized that binding of epinephrine or glucagon to a receptor in the membrane activated synthesis or release of a substance that activated phosphorylation of phosphorylase. This crucial substance was eventually shown (in 1956) to be **adenosine 3′,5′-monophosphate,** now known as **cyclic AMP,** denoted **cAMP** (Figure 37.3).

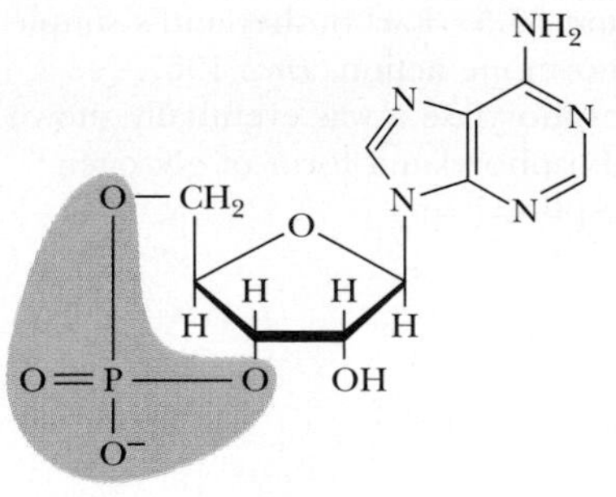

Synthesis and Degradation of Cyclic AMP

Cyclic AMP is produced by an integral membrane enzyme, **adenylyl cyclase** (Figure 37.4). The strain and bond distortion inherent in the bicyclic ring structure of cAMP make its formation an endergonic process, but spontaneous hydrolysis of pyrophosphate formed in this reaction drives the synthesis of cAMP forward. The **phosphodiesterase** reaction hydrolyzes cAMP to AMP as shown. Since this reaction relieves the strain in the cAMP structure, it is highly exergonic ($\Delta G^{\circ\prime} = -50.4$ kJ/mol; see Table 16.1). Sutherland thought of the hormone as the first messenger signaling a need for glycogen breakdown, and he called cAMP a **second messenger.** The basic process, as Sutherland understood it in the 1960s, is shown in Figure 37.5. In this second messenger model, the hormone causes a variety of intracellular effects without itself entering the cell. It was eventually shown that many hormonally activated processes use

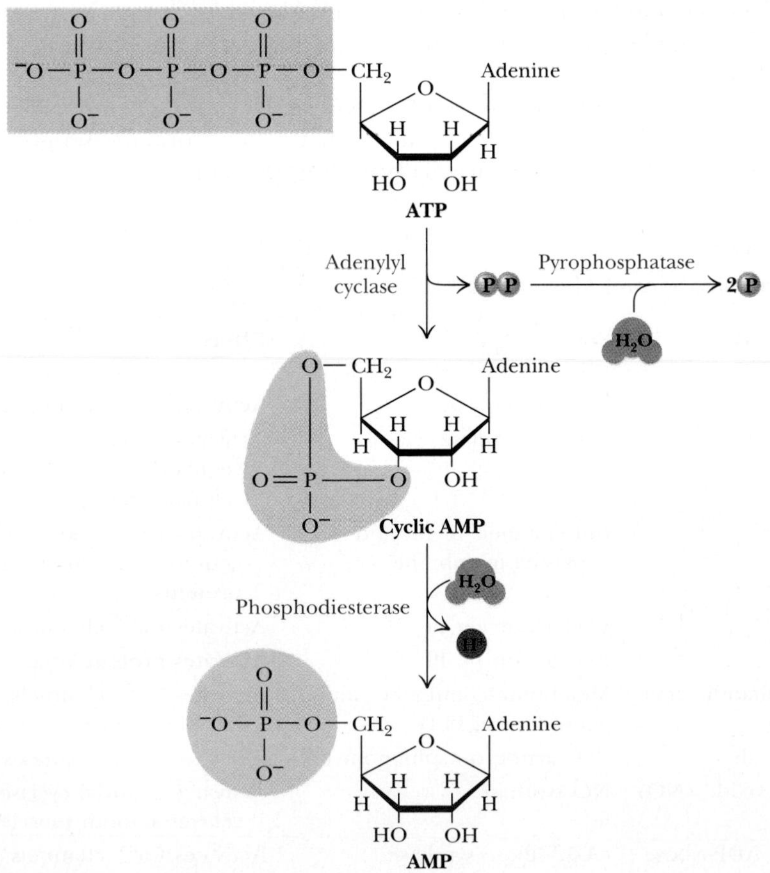

Cyclic AMP

Figure **37.3** The structure of adenosine 3′,5′-cyclic monophosphate.

Figure **37.4** Cyclic AMP is synthesized by membrane-bound adenylyl cyclase and degraded by cytosolic phosphodiesterase.

Figure **37.5**   Earl Sutherland's simple model of hormone action, *circa* 1967. Phosphorylase a was eventually shown to be a phosphorylated form of glycogen phosphorylase.

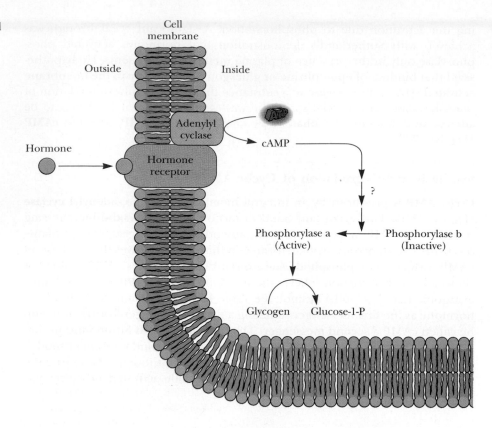

cAMP as a second messenger. For his remarkable achievements in the elucidation of this model, Earl Sutherland received the Nobel Prize in physiology or medicine in 1971.

Since Sutherland's discovery of cAMP, many other second messengers have been identified (Table 37.1). The mediators of second messenger formation for all 7-TMS receptors are GTP-binding proteins.

Table **37.1**

Intracellular Second Messengers*

Messenger	Source	Effect
cAMP	Adenylyl cyclase	Activates protein kinases
cGMP	Guanylyl cyclase	Activates protein kinases, regulates ion channels, regulates phosphodiesterases
Ca^{2+}	Ion channels in ER and plasma membrane	Activates protein kinases, activates Ca^{2+}-modulated proteins
IP_3	PLC action on PI	Activates Ca^{2+} channels
DAG	PLC action on PI	Activates protein kinase C
Phosphatidic acid	Membrane component and product of PLD	Activates Ca^{2+} channels, inhibits adenylyl cyclase
Ceramide	PLC action on sphingomyelin	Activates protein kinases
Nitric oxide (NO)	NO synthase	Activates guanylyl cyclase, relaxes smooth muscle
Cyclic ADP-ribose	cADP-ribose synthase	Activates Ca^{2+} channels

*IP_3 is inositol-1,4,5-trisphosphate; PLC is phospholipase C; PLD is phospholipase D; PI is phosphatidylinositol; DAG is diacylglycerol.

Figure 37.6 The structure of guanylimidodiphosphate.

37.4 GTP-Binding Proteins: The Hormonal Missing Link

Two observations in the early 1970s implicated another protein in the hormonal activation of adenylyl cyclase. First, purification of adenylyl cyclase and the hormone receptor resulted in a loss of hormone stimulation of cyclase activity. Second, Martin Rodbell and his colleagues showed that **GTP** was necessary for hormone activation of adenylyl cyclase. Interestingly, **5′-guanylylimidodiphosphate (GMP-PNP),** a nonhydrolyzable analog of GTP (Figure 37.6), was a "superactivator" of adenylyl cyclase, giving higher cyclase activities than GTP itself. This prompted Rodbell to suggest that the GTP-binding site was the active site of a GTPase. In 1977, Elliott Ross and Alfred Gilman at the University of Virginia reported the partial purification of a GTP-binding protein, which, when reconstituted with the cyclase and hormone receptor, restored hormone stimulation to the adenylyl cyclase reaction. Thus, adenylyl cyclase is not directly activated by the hormone-receptor complex. Instead, binding of hormone to the receptor stimulates a GTP-binding protein (abbreviated now to **G protein**), which in turn activates adenylyl cyclase.

G Proteins

Typically, G proteins are heterotrimers consisting of α- (45 to 47 kD), β- (35 kD), and γ- (7 to 9 kD) subunits. The α-subunit binds GDP or GTP and has an intrinsic, slow GTPase activity. The $G_{\alpha\beta\gamma}$ complex in the unactivated state has GDP at the nucleotide site (Figure 37.7). Binding of hormone to receptor stimulates a rapid exchange of GTP for GDP on G_α. The binding of GTP causes G_α to dissociate from $G_{\beta\gamma}$ and to associate with an effector protein such as adenylyl cyclase. *Binding of G_α (GTP) activates adenylyl cyclase.* The adenylyl cyclase actively synthesizes cAMP as long as G_α(GTP) remains bound to it. However, the intrinsic GTPase activity of G_α eventually hydrolyzes GTP to GDP, leading to dissociation of G_α(GDP) from adenylyl cyclase and reassociation with the $G_{\beta\gamma}$ dimer, regenerating the inactive heterotrimeric $G_{\alpha\beta\gamma}$ complex.

Two stages of amplification occur in the G-protein-mediated hormone response. First, a single hormone-receptor complex can activate many G proteins before the hormone dissociates from the receptor. Second, and more

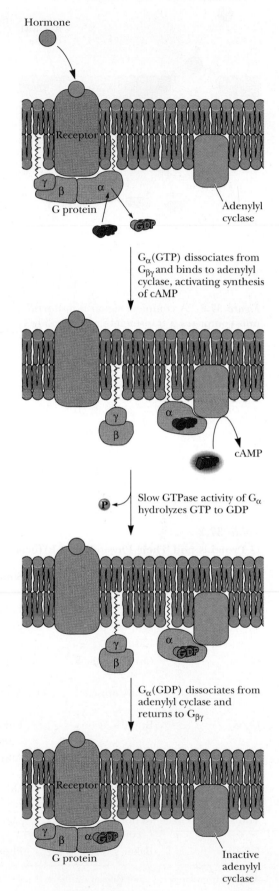

G_α(GTP) dissociates from $G_{\beta\gamma}$ and binds to adenylyl cyclase, activating synthesis of cAMP

Slow GTPase activity of G_α hydrolyzes GTP to GDP

G_α(GDP) dissociates from adenylyl cyclase and returns to $G_{\beta\gamma}$

Figure 37.7 Activation of adenylyl cyclase by heterotrimeric G proteins. Binding of hormone to its receptor causes a conformational change that induces the receptor to catalyze a replacement of GDP by GTP on G_α. The G_α(GTP) complex dissociates from $G_{\beta\gamma}$ and binds to adenylyl cyclase, stimulating synthesis of cAMP. Bound GTP is slowly hydrolyzed to GDP by the intrinsic GTPase activity of G_α. G_α(GDP) dissociates from adenylyl cyclase and reassociates with $G_{\beta\gamma}$. G_α and G_γ are lipid-anchored proteins.

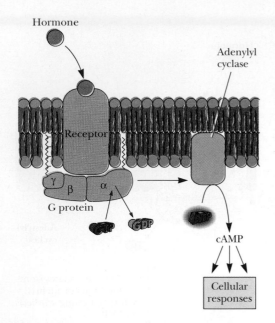

***Figure* 37.8** A complete signal transduction unit comprises a transmembrane signaling pathway consisting of hormone, receptor, and an associated transduction system (which in this case includes a G protein, adenylyl cyclase, and the associated nucleotides).

obvious, the G_α-activated adenylyl cyclase synthesizes many cAMP molecules. Thus, binding of hormone to a very small number of membrane receptors stimulates a large increase in concentration of cAMP within the cell. The hormone receptor, G protein, and cyclase constitute a complete hormone **signal transduction unit** (Figure 37.8).

A given G protein can be activated by several different hormone-receptor complexes. For example, either glucagon or epinephrine, binding to their distinctive receptor proteins, can activate the same species of G protein in liver cells. The effects are additive, and stimulation by glucagon and epinephrine together leads to higher cytoplasmic concentrations of cAMP than activation by either hormone alone.

G proteins are a universal means of signal transduction in higher organisms, activating many hormone-receptor-initiated cellular processes in addition to adenylyl cyclase. Such processes include, but are not limited to, activation of phospholipases C and A_2, and the opening or closing of transmembrane channels for K^+, Na^+, and Ca^{2+} in brain, muscle, heart, and other organs (Table 37.2). G proteins are integral components of sensory pathways such as vision and olfaction. More than 100 different G protein-coupled receptors and at least 21 distinct G proteins are known. At least a dozen different G protein effectors have been identified, including a variety of enzymes and ion channels.

Stimulatory and Inhibitory G Protein Effects

Hormone-receptor-mediated processes regulated by G proteins may be stimulatory (as we have seen) or inhibitory. Each hormone receptor protein interacts specifically with either a stimulatory G protein, denoted $\mathbf{G_s,}$ or with an inhibitory G protein, denoted $\mathbf{G_i}$. For example, the catecholamine hormones (such as epinephrine) bind to four different types of **adrenergic receptor** proteins, designated α_1, α_2, β_1, or β_2 (Chapter 38). Binding to α_1 receptors has no effect on adenylyl cyclase, whereas binding to β_1 or β_2 receptors is

Table **37.2**

G Proteins and Their Physiological Effects

G Protein	Location	Stimulus	Effector	Effect
G_s	Liver	Epinephrine, glucagon	Adenylyl cyclase	Glycogen breakdown
G_s	Adipose tissue	Epinephrine, glucagon	Adenylyl cyclase	Fat breakdown
G_s	Kidney	Antidiuretic hormone	Adenylyl cyclase	Conservation of water
G_s	Ovarian follicle	Luteinizing hormone	Adenylyl cyclase	Increased estrogen and progesterone synthesis
G_i	Heart muscle	Acetylcholine	Potassium channel	Decreased heart rate and pumping force
G_i/G_o	Brain neurons	Enkephalins, endorphins, opioids	Adenylyl cyclase, potassium channels, calcium channels	Changes in neuron electrical activity
G_q	Smooth muscle cells in blood vessels	Angiotensin	Phospholipase C	Muscle contraction, blood pressure elevation
G_{olf}	Neuroepithelial cells in the nose	Odorant molecules	Adenylyl cyclase	Odorant detection
Transducin (G_t)	Retinal rod and cone cells	Light	cGMP phosphodiesterase	Visual signal detection
GPA1	Baker's yeast	Pheromones	Unknown	Mating

Adapted from Hepler, J., and Gilman, A., 1992. G proteins. *Trends in Biochemical Sciences* **17:**383–387.

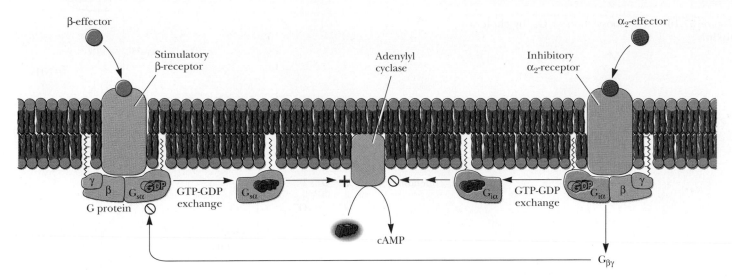

Figure 37.9 Adenylyl cyclase activity is modulated by the interplay of stimulatory (G_s) and inhibitory (G_i) G proteins. Binding of hormones to β_1- and β_2-adrenergic receptors activates adenylyl cyclase via G_s, whereas hormone binding to α_2 receptors leads to the inhibition of adenylyl cyclase. Inhibition may occur by direct inhibition of cyclase activity by $G_{i\alpha}$ or by binding of $G_{i\beta\gamma}$ to $G_{s\alpha}(GTP)$.

stimulatory and binding to α_2 receptors is inhibitory. The β_1 and β_2 receptors interact specifically with a G_s complex, and α_2 receptors interact only with G_i proteins. The "G protein" described in Figure 37.7 is in fact a G_s-type G protein. Two possibilities exist (Figure 37.9). Binding of a hormone to its receptor triggers GTP $\rightleftharpoons$ GDP exchange and dissociation of $G_{i\alpha}(GTP)$ from $G_{i\beta\gamma}$. Inhibition may then occur *either* by binding of $G_{i\alpha}(GTP)$ to adenylyl cyclase to directly inhibit the cyclase, *or* by action of $G_{i\beta\gamma}$, which can compete with the cyclase for $G_{s\alpha}(GTP)$ complexes. The presence in liver cell membranes of much more G_i than G_s favors the competitive role for $G_{i\beta\gamma}$.

G_s Is the Site of Action of Cholera Toxin

Many of the details of G protein mediation of hormonal effects have been elucidated through studies of the potent effects of bacterial toxins, including **cholera toxin** and **pertussis toxin.** *Vibrio cholerae,* the Gram-negative bacterium that causes cholera, induces severe diarrhea in its victims, leading to death if the fluids are not replenished. Cholera toxin is an 87-kD protein consisting of an A subunit and five B subunits. The B subunits act in host-cell recognition. The A subunit consists of A_1 and A_2 peptides linked by a disulfide bond. The 22-kD A_1 peptide catalyzes the *ADP-ribosylation* of Arg^{201} in the α-subunit of G_s (Figure 37.10), using NAD^+ as a substrate. ADP-ribosylation strongly inhibits the GTPase activity of $G_{s\alpha}$, effectively trapping $G_{s\alpha}$ in the activated state and causing prolonged activation of adenylyl cyclase. Elevated levels of cyclic AMP in turn cause intestinal epithelial cells to secrete high volumes of fluid. Both the bacteria and the cholera toxin remain localized in the intestines through the course of the disease, but if fluids are actively replaced, the body's immune system eventually gains control and destroys the bacteria.

Pertussis Toxin Causes ADP-Ribosylation of G_i

ADP-ribosylation of a Cys residue on $G_{i\alpha}$ is catalyzed by pertussis toxin, a product of *Bordetella pertussis,* the bacterium that causes **whooping cough.** Pertussis toxin is a 110-kD hexameric protein (a 28-kD A subunit and five B-like

Figure 37.10 ADP-ribosylation of $G_{s\alpha}$ by cholera toxin.

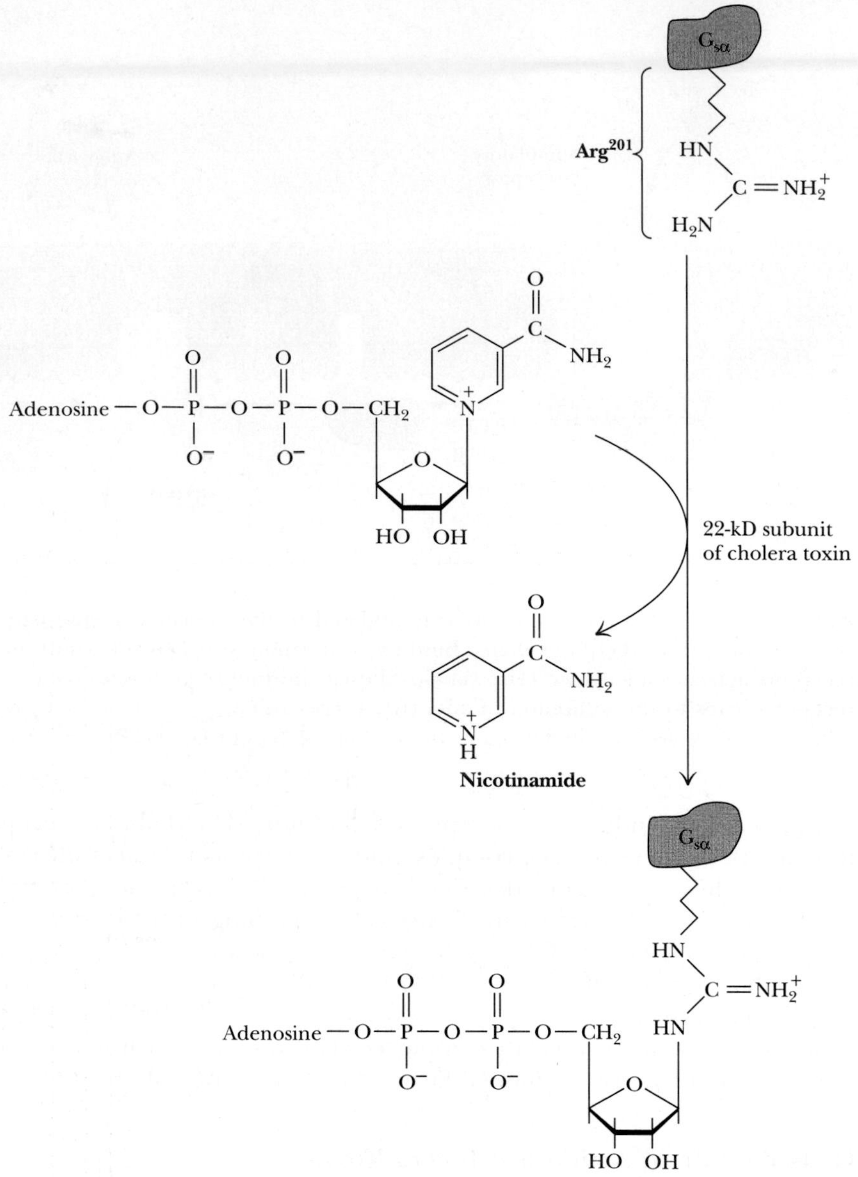

subunits.) In this case, the ADP-ribosylation inhibits exchange of GDP for GTP, thereby preventing $G_{i\alpha}$ from inhibiting adenylyl cyclase. Pertussis, in contrast to cholera, is a systemic infection, and the breakdown of adenylyl cyclase regulation is felt by tissues throughout the body.

Fluoride Ion Stimulation of Adenylyl Cyclase

Early studies of adenylyl cyclase revealed that cyclase activity was stimulated by fluoride ion, F^-, but the nature of this activation remained a mystery for some 20 years. The stimulation of adenylyl cyclase by F^- is in fact due to action on G_s. The fluoride effect depends upon trace contamination of laboratory solutions, reagents, and glassware with aluminum ion, Al^{3+}. Aluminum has a high affinity for F^-, and forms several complexes such as AlF_3 and AlF_4^-. One of these complexes (it is not known with certainty which is responsible) binds together with GDP at the nucleotide site of $G_{s\alpha}$, with the aluminum fluoride complex presumably located at the site normally occupied by the terminal phosphate of GTP (Figure 37.11). The GDP–aluminum fluoride complex resembles GTP and behaves in a manner similar to GTP itself, activating $G_{s\alpha}$, which stimulates adenylyl cyclase in the usual way.

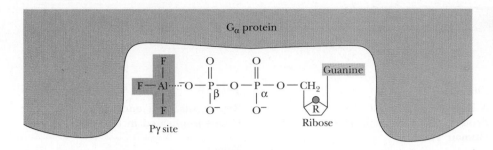

***Figure* 37.11** Fluoride activation of adenylyl cyclase arises from the coordination of an aluminum fluoride complex (most likely AlF$_3$) together with GDP at the GTP site of G$_{s\alpha}$. The aluminum fluoride complex mimics the γ-phosphoryl group of GTP.

ras and the Small GTP-Binding Proteins

GTP-binding proteins are implicated in critical growth mechanisms in higher organisms. Certain tumor virus genomes contain genes encoding 21-kD proteins that bind GTP and show regions of homology with other G proteins. The first of these genes to be identified was found in *rat* *s*arcoma virus, and was dubbed the **ras gene.** Genes implicated in tumor formation are known as **oncogenes;** often they are mutated versions of normal, noncancerous genes involved in growth regulation, so-called **proto-oncogenes.** The normal, cellular *ras* protein is a GTP-binding protein that functions similarly to other G proteins described above, activating metabolic processes when GTP is bound and becoming inactive when GTP is hydrolyzed to GDP. The GTPase activity of the normal *ras* p21 is very low, as appropriate for a G protein that regulates long-term effects like growth and differentiation. A specific **GTPase-activating protein (GAP)** increases the GTPase activity of the *ras* protein. Mutant (oncogenic) *ras* proteins have severely impaired GTPase activity, which apparently causes serious alterations of cellular growth and metabolism in tumor cells. Several crystallographic studies have determined the structures of *ras* proteins (Figure 37.12) in complexes with GDP, GMP-PNP, and a photolabile GTP derivative. Comparisons of these structures reveal that the most dramatic differences between the GDP and GTP complexes of *ras* are in the region (residues 32 to 36) that is thought to interact with GAP. This conformation change causes a change in coordination of a bound Mg^{2+} ion. In the *ras*-GTP complex, Mg^{2+} coordinates the β- and γ-phosphates of bound GTP, but, in the *ras*-GDP complex, the interaction of Mg^{2+} with the (now missing) γ-P is replaced by coordination with the carboxyl group of Asp^{57}.

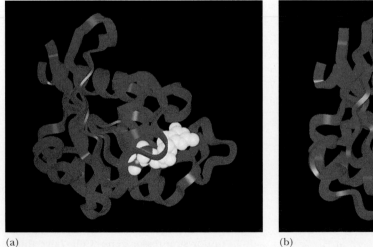 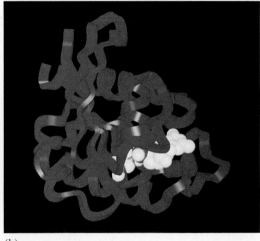

(a)   (b)

***Figure* 37.12** The structure of the *ras* p21 complexed with (a) GDP and (b) GMP-PNP. The *ras* p21-GMP-PNP complex is the active conformation of this protein. Inside the front cover of this book is an illustration that includes a signal transduction pathway involving *ras*.

A Deeper Look

Cancer, Oncogenes, and Tumor Suppressor Genes

The disease state known as **cancer** is the uncontrolled growth and proliferation of one or more cell types in the body. Control of cell growth and division is an incredibly complex process, involving the signal-transducing proteins (and small molecules) described in this chapter and many others like them. The genes that give rise to these growth-controlling proteins are of two distinct types:

1. **Oncogenes:** These genes code for proteins that are capable of stimulating cell growth and division. In normal tissues and organisms, such growth-stimulating proteins are regulated, so that growth is appropriately limited. However, mutations in these genes may result in loss of growth regulation, leading to uncontrolled cell proliferation and tumor development. These mutant genes are known as *oncogenes,* because they induce the oncogenic state—cancer. The normal precursors of these genes are termed **proto-oncogenes** and are essential for normal cell growth and differentiation. Oncogenes are *dominant,* because a mutation of only one of the cell's two copies of that gene can lead to tumor formation. Table A lists a few of the known oncogenes (over 60 are now known).

2. **Tumor suppressor genes:** These genes code for proteins whose normal function is to *turn off* cell growth. A mutation in one of these growth-limiting genes may result in a protein product that has lost its growth-limiting ability. The normal forms of such genes have been shown to suppress tumor

Table A

A Representative List of Proto-Oncogenes Implicated in Human Tumors

Proto-Oncogene	Neoplasm(s)
Abl	Chronic myelogenous leukemia
ErbB-1	Squamous cell carcinoma; astrocytoma
ErbB-2 (Neu)	Adenocarcinoma of breast, ovary, and stomach
Gip	Carcinoma of ovary and adrenal gland
Gsp	Adenoma of pituitary gland; carcinoma of thyroid
Myc	Burkitt's lymphoma; carcinoma of lung, breast, and cervix
L-Myc	Carcinoma of lung
N-Myc	Neuroblastoma; small cell carcinoma of lung
H-Ras	Carcinoma of colon, lung, and pancreas; melanoma
K-Ras	Acute myelogenous and lymphoblastic leukemia; carcinoma of thyroid; melanoma
N-Ras	Carcinoma of genitourinary tract and thyroid; melanoma
Ret	Carcinoma of thyroid
Ros	Astrocytoma
K-Sam	Carcinoma of stomach
Sis	Astrocytoma
Src	Carcinoma of colon
Trk	Carcinoma of thyroid
Jun Fos	} Several

Adapted from Bishop, J. M., 1991. Molecular themes in oncogenesis. *Cell* **64**:235–248.

Table B

Representative Tumor Suppressor Genes Implicated in Human Tumors

Tumor Suppressor Gene	Neoplasm(s)
RB1	Retinoblastoma; osteosarcoma; carcinoma of breast, bladder, and lung
P53	Astrocytoma; carcinoma of breast, colon, and lung; osteosarcoma
WT1	Wilms' tumor
DCC	Carcinoma of colon
NF1	Neurofibromatosis type 1
FAP	Carcinoma of colon
MEN-1	Tumors of parathyroid, pancreas, pituitary, and adrenal cortex

Adapted from Bishop, J. M., 1991. Molecular themes in oncogenesis. *Cell* **64**:235–248.

growth and are known as *tumor suppressor genes.* Since both cellular copies of a tumor suppressor gene must be mutated to foil its growth-limiting action, these genes are *recessive* in nature. Table B presents several recognized tumor suppressor genes.

Careful molecular analysis of cancerous tissue has shown that tumor development may result from mutations in several proto-oncogenes or tumor suppressor genes. The implication is that *there is redundancy in cellular growth regulation.* Many (if not all) tumors are either the result of interactions of two or more oncogene products or arise from simultaneous mutations in a proto-oncogene and both copies of a tumor suppressor gene. Cells have thus evolved with overlapping growth-control mechanisms. When one is compromised by mutation, others take over.

37.5 The 7-TMS Receptors

The primary and secondary structures of the 7-transmembrane segment (7-TMS) receptors are similar to those of bacteriorhodopsin (see Chapters 9, 35) and rhodopsin (see Chapter 38). The **α- and β-adrenergic receptors,** for which epinephrine is a ligand, are good examples (Figure 37.13). Hydropathy analysis of these receptors is consistent with seven transmembrane α-helical segments. The extracellular N-terminal segment has two glycosylation sites. The hydrophilic loops connecting the seven hydrophobic domains are not required for ligand binding. Instead, the ligand-binding site (for the cationic catecholamines) is located within the hydrophobic core of the receptor.

The binding of epinephrine to a β-adrenergic receptor initiates the above-described G protein activation of adenylyl cyclase. α_1-Adrenergic receptors stimulate inositol phospholipid metabolism when activated (see following section). Stimulation of α_2-adrenergic receptors appears to counteract hormone-stimulated increases in [cAMP]. The β-adrenergic receptors act through G proteins. These G proteins are coupled to several pathways, including adenylyl and guanylyl cyclases, phospholipases A and C, Ca^{2+} and K^+ channels, and phosphodiesterases. Substitution of Asp^{113} in the third hydrophobic domain of the β-adrenergic receptor with an Asn or Gln by site-directed mutagenesis results in a dramatic decrease in affinity of the receptor for both agonists and antagonists. Significantly, this Asp residue is conserved in all other G protein-coupled receptors that bind biogenic amines, but is absent in receptors whose ligands are not amines. Asp^{113} appears to be the counter-ion for the amine moiety of adrenergic ligands.

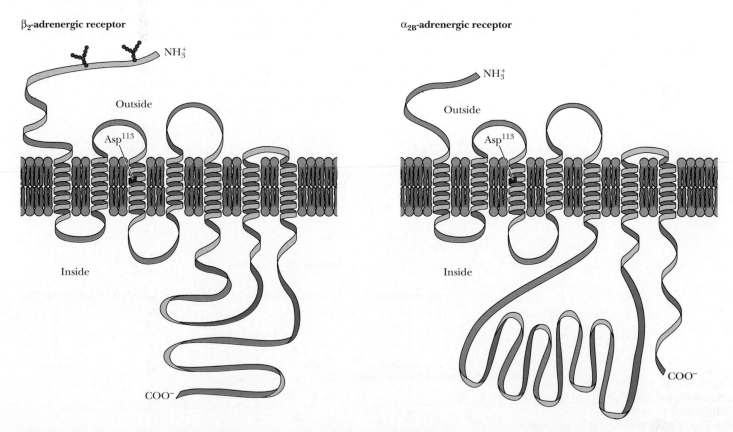

β₂-adrenergic receptor

NH_3^+

Outside

Asp^{113}

Inside

COO^-

α₂ᵦ-adrenergic receptor

NH_3^+

Outside

Asp^{113}

Inside

COO^-

Figure **37.13**   The arrangements of the β_2- and α_2-adrenergic receptors in the membrane.

The β-adrenergic receptor is desensitized by phosphorylation either by a specific **β-adrenergic receptor kinase (βARK)** or by **protein kinase A (PKA),** the cAMP-dependent protein kinase. Phosphorylation sites for both kinases are located in the receptors' C-terminal domain. The sites phosphorylated by protein kinase A are adjacent to segments of the receptor that mediate coupling to G proteins, implying that phosphorylation at these sites interrupts this coupling. The desensitization due to βARK occurs by a different mechanism; it may involve another as yet unidentified protein.

37.6 Specific Phospholipases Release Second Messengers

A diverse array of second messengers are generated by breakdown of membrane phospholipids. Binding of certain hormones and growth factors to their respective receptors triggers a sequence of events that can lead to the activation of **specific phospholipases.** Action of these phospholipases on membrane lipids produces the second messengers shown in Figure 37.14.

Inositol Phospholipid Breakdown Yields Inositol-1,4,5-Trisphosphate and Diacylglycerol

Breakdown of **phosphatidylinositol (PI)** and its derivatives by **phospholipase C** produces a family of second messengers. In the best-understood pathway, successive phosphorylations of PI produce **phosphatidylinositol-4-P (PIP)** and **phosphatidylinositol-4,5-bisphosphate (PIP$_2$)** (Figure 37.15). Four isozymes of phospholipase C (denoted α, β, γ, and δ) hydrolyze PI, PIP, and PIP$_2$. Hydrolysis of PIP$_2$ by phospholipase C yields the second messenger **inositol-1,4,5-trisphosphate (IP$_3$),** as well as another second messenger, **diacylglycerol (DAG).** IP$_3$ is water-soluble and diffuses to intracellular organelles where release of Ca^{2+} is activated. DAG, on the other hand, is lipophilic and remains in the plasma membrane where it activates a Ca^{2+}-dependent protein kinase known as **protein kinase C** (see following discussion).

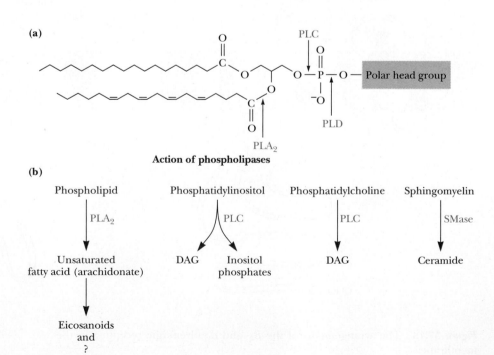

Figure 37.14 (a) The general action of phospholipase A$_2$ (PLA$_2$), phospholipase C (PLC), and phospholipase D (PLD). (b) The synthesis of second messengers from phospholipids by the action of phospholipases and sphingomyelinase.

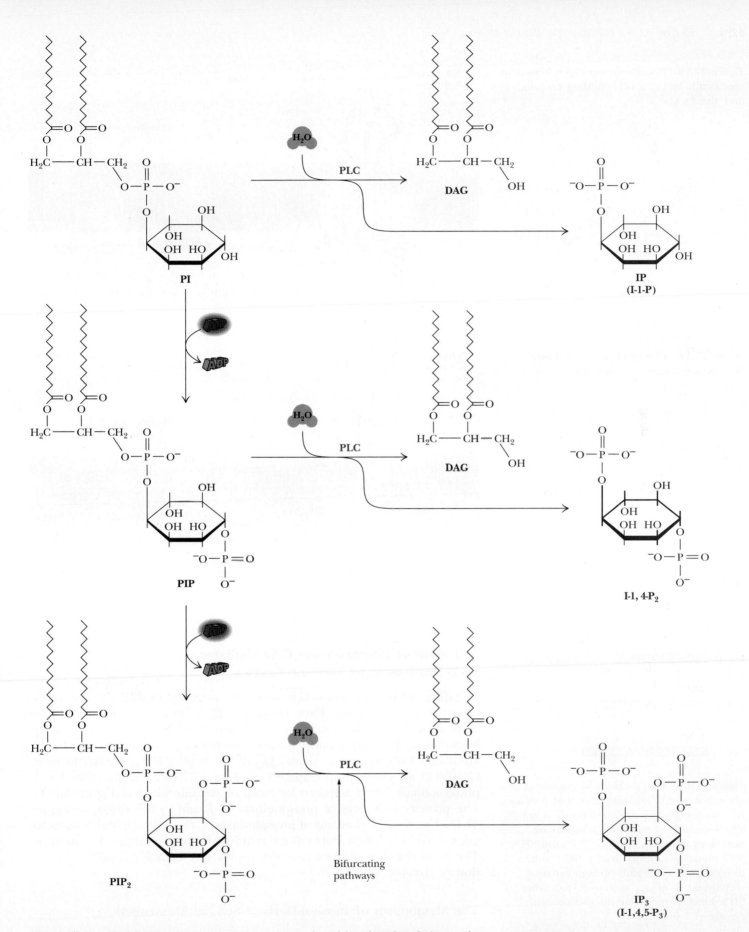

***Figure* 37.15** The family of second messengers produced by phosphorylation and breakdown of phosphatidylinositol. PLC action instigates a bifurcating pathway culminating in two distinct and independent second messengers, as in DAG and IP₃.

Figure 37.16 Phospholipase C-β is activated specifically by G_q, a GTP-binding protein, and also by Ca^{2+}.

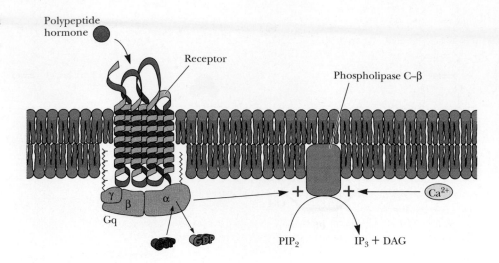

Figure 37.17 Phospholipase C-γ is activated by receptor tyrosine kinases and by Ca^{2+}.

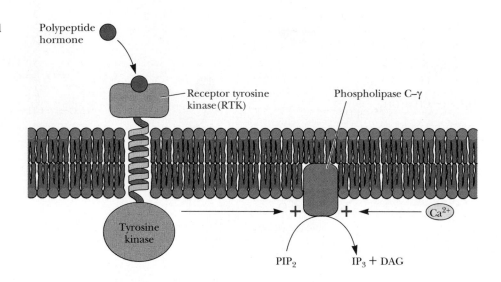

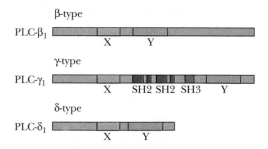

Figure 37.18 The amino acid sequences of phospholipase C isozymes β, γ, and δ share two homologous domains, denoted X and Y. The sequence of γ-isozyme contains ***src* homology domains,** denoted SH2 and SH3. SH2 domains (approximately 100 residues in length) interact with phosphotyrosine-containing proteins (such as RTKs), whereas SH3 domains modulate interactions with cytoskeletal proteins.

(Adapted from Dennis, E., Rhee, S., Billah, M., and Hannun, Y., 1991. Role of phospholipases in generating lipid second messengers in signal transduction. The FASEB Journal 5:2068–2077.)

Activation of Phospholipase C Is Mediated by G Proteins or by Tyrosine Kinases

The different phospholipase C isozymes are activated by different intracellular events (Figure 37.16). Phospholipase C-β, -γ, and -δ are all Ca^{2+}-dependent. In addition, phospholipase C-β is stimulated by a class of G proteins known as G_q. Binding of polypeptide hormones such as vasopressin or bradykinin to 7-TMS receptors releases $G_{q\alpha}(GTP)$ from a $G_{q\alpha\beta\gamma}$ trimer. In turn, $G_{q\alpha}(GTP)$ activates phospholipase C-β (Figure 37.16). On the other hand, phospholipase C-γ is activated by **receptor tyrosine kinases** (Figure 37.17). The primary structures of phospholipase C-β and -γ are shown in Figure 37.18. The X and Y domains of phospholipase C-β and -γ are highly homologous, and both of these domains are required for phospholipase C activation. The other domains of these isozymes confer specificity for G protein activation or tyrosine kinase activation.

The Metabolism of Inositol-Derived Second Messengers

A summary of PIP_2-related signaling processes is shown in Figure 37.19. IP_3 has a cellular half-life of only a few seconds. It is rapidly processed along two principal paths: (a) it can be catabolized by a series of phosphatases to yield

A Deeper Look

PI Metabolism and the Pharmacology of Li$^+$

An intriguing aspect of the phospho-inositide story is the specific action of lithium ion, Li$^+$, on several steps of PI metabolism. Lithium salts have been used in the treatment of *manic-depressive illnesses* for more than 30 years, but the mechanism of lithium's therapeutic effects had been unclear. Recently, how-ever, several of the dephosphorylation reactions in Figure 37.19 have been shown to be sensitive to Li$^+$ ion. Li$^+$ levels similar to those employed in treatment of manic illness thus lead to the accumulation of several key intermediates, including I-1,3,4-P$_3$, I-1,4-P$_2$, I-3-P, and I-4-P. In addition, *myo-inositol-1-phosphatase*, another enzyme involved in phosphoinositide metabolism, is inhibited by Li$^+$ with a K_I of 1 mM. This story is far from complete, and many new insights into phospho-inositide metabolism and the effects of Li$^+$ can be anticipated.

inositol-1,4-bisphosphate, inositol-4-phosphate, and free *myo*-inositol, which can be subsequently reincorporated into *new* inositol phospholipids; or (b) it can be phosphorylated to yield inositol-1,3,4,5-tetraphosphate, which then undergoes a complex series of phosphorylations and dephosphorylations to at least six other inositol phosphate compounds. *Interestingly, many of these inositol phosphates have also been shown to behave as second messengers in various cellular processes.* Some of these compounds may serve as extracellular signals in the regulation of specific neural mechanisms.

Figure 37.19 The pathways of phosphoinositide biosynthesis and metabolism. The Li$^+$-sensitive steps are shown as red lines.

(Adapted from Catt, K., and Balla, T., 1989. Phosphoinositide metabolism and hormone action. Annual Review of Medicine 40:487–509.)

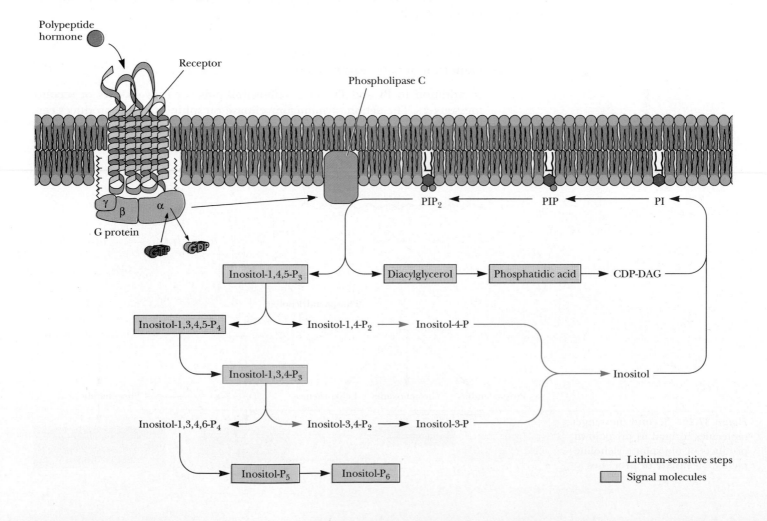

Second Messengers Are Derived from Phosphatidylcholine Breakdown

The second messenger effects of cAMP, Ca^{2+}, and the products of PI metabolism are the best understood, but other pathways are also known. For example, **phosphatidylcholine breakdown** occurs in a variety of cellular signaling pathways. Hydrolysis of phosphatidylcholine (PC) by phospholipase A_2 is an important source of **arachidonic acid,** the precursor of **eicosanoids** (Chapter 24). However, PC can also be hydrolyzed by phospholipases C and D to yield DAG and phosphatidic acid, respectively (Figure 37.20). These hydrolyses are stimulated by hormones, by G proteins, and by Ca^{2+} (via IP_3) and protein kinase C (via DAG). In at least some cases, the phospholipases C and D acting here are specific for PC and distinct from the phospholipases that act on PI and PIP_2.

The roles of PC breakdown products are not entirely clear, but several potential roles are postulated. The large amount of PC in cell membranes means that phospholipase C action on PC can produce sustained quantities of DAG, which could lead to prolonged activation of protein kinase C. PIP_2, with a cellular concentration much less than 1% that of PC, yields only small amounts of DAG for short times. Moreover, hydrolysis of PC is selective, since it can produce DAG (to activate protein kinase C) without mobilization of intracellular Ca^{2+} (via IP_3 production). Hydrolysis of PC by phospholipase D to produce phosphatidic acid (PA) may also represent a significant signaling pathway. PA promotes Ca^{2+} entry into cells and also mobilizes intracellular Ca^{2+}. PA also inhibits adenylyl cyclase by interacting with G_i. Activation by PA of PIP_2-dependent phospholipase C has been observed both *in vitro* and *in vivo*. PA also activates muscle phosphorylase kinase, as well as DNA synthesis in certain cells.

Sphingomyelin and Glycosphingolipids Also Generate Second Messengers

In addition to PI and PC, other phospholipids serve as sources of second messengers. The action of **sphingomyelinase** on sphingomyelin produces **ceramide,** which stimulates **ceramide-activated protein kinase.** Similarly, ganglio-

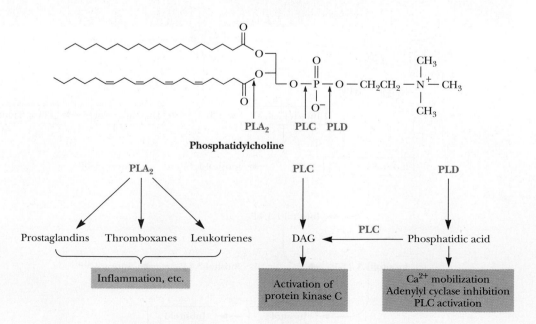

***Figure* 37.20** Second messenger molecules formed as breakdown products of phosphatidylcholine mediate a variety of processes.

sides (such as ganglioside G_{M3} (Chapter 9)) and their breakdown products modulate the activity of protein kinases and G protein-coupled receptors.

37.7 Calcium as a Second Messenger

Calcium ion is an important intracellular signal. Binding of certain hormones and signal molecules to plasma membrane receptors can cause transient increases in cytoplasmic Ca^{2+} levels, which in turn can activate a wide variety of enzymatic processes, including smooth muscle contraction, exocytosis, and glycogen metabolism. (Most of these activation processes depend on special Ca^{2+}-binding proteins discussed in the following section.) Cytoplasmic $[Ca^{2+}]$ can be increased in two ways (Figure 37.21). As mentioned briefly above, cAMP can activate the opening of plasma membrane Ca^{2+} channels, allowing extracellular Ca^{2+} to stream in. On the other hand, cells also contain intracellular reservoirs of Ca^{2+}, within the endoplasmic reticulum and **calciosomes,** small membrane vesicles that are similar in some ways to muscle sarcoplasmic reticulum. These special intracellular Ca^{2+} stores are *not* released by cAMP. They respond to IP_3, a second messenger derived from **phosphatidylinositol (PI).**

Calcium-Induced Calcium Release

Much of the Ca^{2+} entering the cytoplasm by action of IP_3 appears to come from two sources: (a) parts of the endoplasmic reticulum that are closely associated with the plasma membrane, and (b) the extracellular environment. As shown in Figure 37.22, Ca^{2+} release appears to be a two-step process. IP_3 binding to receptors on the ER membrane opens Ca^{2+} channels releasing Ca^{2+} from the ER. Released Ca^{2+} binds to the ER *IP_3 receptor,* causing a conformational change that induces the opening of adjacent *plasma membrane Ca^{2+} channels.* This conformational coupling between a receptor on one membrane and a Ca^{2+} channel on an adjacent membrane is remarkably similar to the operation of the ryanodine receptor in sarcoplasmic reticulum (Chapter 36), and the IP_3 receptor shares remarkable structural similarities with the ryanodine receptor (Figure 37.23).

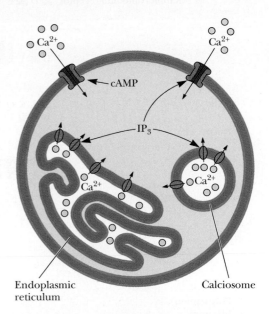

Endoplasmic reticulum   Calciosome

***Figure* 37.21**   Cytoplasmic $[Ca^{2+}]$ increases occur via the opening of Ca^{2+} channels in the membranes of calciosomes, the endoplasmic reticulum, and the plasma membrane.

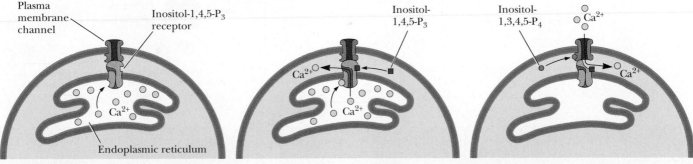

Plasma membrane channel   Inositol-1,4,5-P_3 receptor   Inositol-1,4,5-P_3 Inositol-1,3,4,5-P_4

Endoplasmic reticulum

(a) Resting state

(b) Inositol-1,4,5-P_3 mobilizes calcium from endoplasmic reticulum pool

(c) Emptying of the endoplasmic reticulum pool induces a conformational change in the inositol-1,4,5-P_3 receptor complex, which opens the plasma membrane calcium channel, perhaps in conjunction with inositol-1,3,4,5-P_4

***Figure* 37.22** IP_3 mediates Ca^{2+}-induced Ca^{2+} release. Binding of IP_3 to the ER IP_3 receptor opens ER Ca^{2+} channels. Flow of Ca^{2+} through these channels induces a conformational change that opens plasma membrane Ca^{2+} channels.

(Adapted from Berridge, M., 1990. Calcium oscillations. Journal of Biological Chemistry 265:9583–9586.)

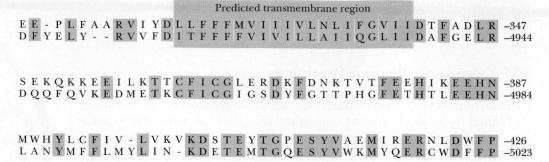

| Mouse IP₃ receptor | E T E Q D K E H T C E T L L M C I V T V L S H G L R S G G G Y G D V L R K P S K | –308 |
| Rabbit ryanodine receptor | E D E P E P D M K C D D M M T C Y L F H M Y V G V R A G G G I G D E I E D P A G | –4906 |

Predicted transmembrane region

| E E - P L F A A R V I Y D L L F F F M V I I I V L N L I F G V I I D T F A D L R | –347 |
| D F Y E L Y - - R V V F D I T F F F F V I V I L L A I I Q G L I I D A F G E L R | –4944 |

| S E K Q K K E E I L K T T C F I C G L E R D K F D N K T V T F E E H I K E E H N | –387 |
| D Q Q F Q V K E D M E T K C F I C G I G S D Y F G T T P H G F E T H T L E E H N | –4984 |

| M W H Y L C F I V - L V K V K D S T E Y T G P E S Y V A E M I R E R N L D W F P | –426 |
| L A N Y M F F L M Y L I N - K D E T E M T G Q E S Y V W K M Y Q E R C W D F F P | –5023 |

***Figure* 37.23**   Sequence homology between the mouse IP₃ receptor and the rabbit muscle ryanodine receptor. Identical residues are shaded blue.

(Adapted from Mignery, G., Sudhof, T., Takel, K., and Camilli, P., 1989. Putative receptor for inositol-1,4,5-trisphosphate is similar to the ryanodine receptor. Nature 342:192–195.)

The actions of the two second messengers IP₃ and DAG are complementary (Figure 37.24). IP₃ elevates cytoplasmic Ca^{2+} levels and DAG activates protein kinase C in a Ca^{2+}- and phosphatidylserine-dependent manner.

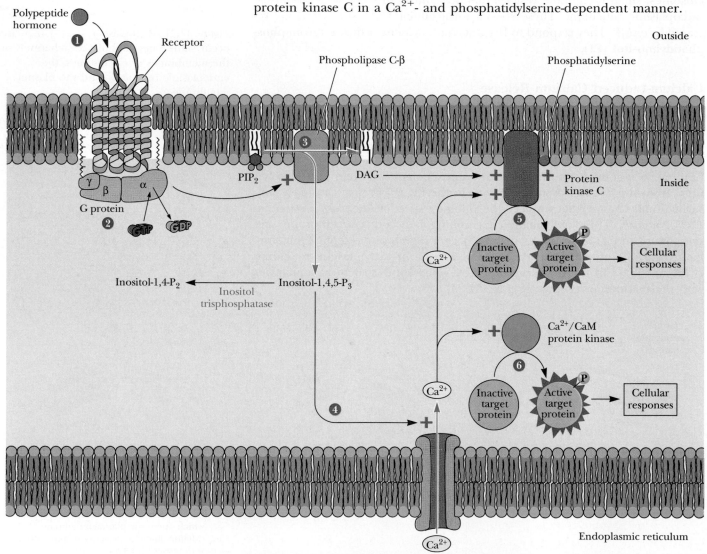

***Figure* 37.24**   IP₃-mediated signal transduction pathways. Increased $[Ca^{2+}]$ activates protein kinases that phosphorylate target proteins. Ca^{2+}/CaM represents calci-calmodulin (Ca^{2+} complexed with the regulatory protein calmodulin).

Calcium Oscillations

One of the most intriguing and exciting developments in the field of Ca^{2+} regulation has been the discovery by Michael Berridge and others that the increases in intracellular $[Ca^{2+}]$ induced by IP_3 are **oscillatory** in nature! Several examples of Ca^{2+} oscillatory behavior are shown in Figure 37.25, including patterns of near-sinusoidal oscillations and baseline spikes. Ca^{2+} oscillations are induced most often by activation of receptors that act through the phosphoinositide pathway. Examples include *α₁-adrenergic* receptors, *vasopressin* and *angiotensin* receptors in liver, *histamine* receptors in endothelial cells, *cholecystokinin* receptors in pancreatic cells, and *B₂ bradykinin* receptors in chromaffin cells.

Several models have been proposed to account for oscillatory behavior, including models involving oscillations in IP_3 and models in which a constant IP_3 level induces fluctuations in the uptake and release of Ca^{2+}. The purpose of Ca^{2+} oscillations in cells is not understood, but two possibilities include (a) the need to protect sensitive intracellular processes from prolonged high levels of Ca^{2+}, and (b) the need to create spatial "waves" of Ca^{2+} in the cell. Many Ca^{2+} oscillatory systems display a spatial organization, so that Ca^{2+} transients such as those of Figure 37.25 actually spread through a cell in the form of a wave of Ca^{2+} that propagates at a rate of 10 to 100 $\mu m/sec$. There is also evidence that Ca^{2+} waves may spread from one cell to the next. For example, activated cells in the ciliated epithelium of the lung excite neighboring cells through a signal, thought to be Ca^{2+}, that radiates outward in a wavelike manner at approximately 10 $\mu m/sec$.

Intracellular Calcium-Binding Proteins

Given the central importance of Ca^{2+} as an intracellular messenger, it should not be surprising that complex mechanisms exist in cells to manage and control Ca^{2+}. When Ca^{2+} signals are generated by cAMP, IP_3, and other agents, these signals are translated into the desired intracellular responses by **calcium-binding proteins,** which in turn regulate many cellular processes. One of these, protein kinase C, will be described in Section 37.8. The other important Ca^{2+}-binding proteins can, for the most part, be divided into two groups on the basis of structure and function: (a) the **calcium-modulated proteins,** including **calmodulin, parvalbumin, troponin C,** and many others, all of which have in common a structural feature called the **EF hand** (Figure 5.25 and Figure 36.23), and (b) the **annexin proteins,** a family of homologous proteins that interact with membranes and phospholipids in a Ca^{2+}-dependent manner.

More than 170 calcium-modulated proteins are known (Table 37.3). All possess a characteristic peptide domain consisting of a short α-helix, a loop of 12 amino acids, and a second α-helix. Robert Kretsinger at the University of Virginia initially discovered this pattern in **parvalbumin,** a protein first identified in the carp fish and later in neurons possessing a high firing rate and a high oxidative metabolism. Kretsinger lettered the six helices of parvalbumin A through F. He noticed that the E and F helices, joined by a loop, resembled the thumb and forefinger of a right hand (Figure 5.25), and named this structure the *EF hand,* a name in common use today to identify the helix-loop-helix motif in calcium-binding proteins. In the EF hand, Ca^{2+} is coordinated by six carboxyl oxygens contributed by a glutamate and three aspartates, by a carbonyl oxygen from a peptide bond, and by the oxygen of a coordinated water molecule. The EF hand was subsequently identified in *calmodulin, troponin C,* and **calbindin-9K** (Figure 37.26). Most of the known EF-hand proteins

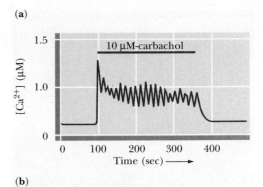

(a)

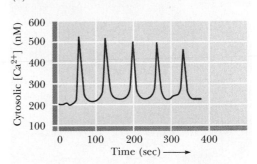

(b)

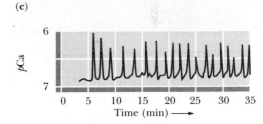

(c)

***Figure* 37.25** Induced oscillations of cytoplasmic $[Ca^{2+}]$. (a) Carbachol-dependent oscillations in the parotid gland. (b) Response of hepatocytes to norepinephrine. (c) Endothelial cell response to histamine.

(Adapted from Berridge, M., 1990. Calcium oscillations. Journal of Biological Chemistry 265:9583–9586.)

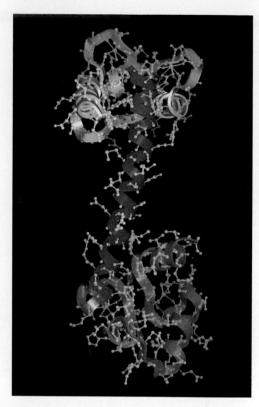

Figure 37.26 Structure of calmodulin. Calmodulin, with four Ca^{2+}-binding domains, forms a dumbbell-shaped structure with two globular domains joined by an extended, central helix. Each globular domain juxtaposes 2 Ca^{2+}-binding EF-hand domains. An intriguing feature of these EF-hand domains is their nearly identical three-dimensional structure, despite a relatively low degree of sequence homology (only 25% in some cases).

Table 37.3
Some Calcium-Modulated Proteins

Protein	Function
α-Actinin	Cross-linking of cytoskeletal F-actin
Calcineurin B	Protein Ser/Thr phosphatase
Calmodulin	Modulates activity of Ca^{2+}-dependent proteins
Calretinin	Modulates Ca^{2+}-dependent neural processes
Caltractin	Modulates Ca^{2+}-sensitive contractile fibers
β- and γ-Crystallins	Ca^{2+}-modulated processes in eye lens
Flagellar Ca^{2+}-binding protein	Flagellar function and cell motility
Frequinin	Phototransduction in retinal cone cells
Inositol phospholipid-specific phospholipase C	Second messenger release and cell signaling
Myeloperoxidase	Inflammatory action of neutrophils
Parvalbumin	Acceleration of muscle relaxation, Ca^{2+} sequestration
S-100	Cell cycle progression, cell differentiation, cytoskeleton-membrane interactions
Thioredoxin reductase	Electron transfer processes in keratinocytes
Troponin C	Activation of muscle contraction

Adapted from Heizmann, C. W., ed., 1991. *Novel Calcium Binding Proteins—Fundamentals and Clinical Implications.* New York: Springer-Verlag.

possess two or more (as many as eight) EF-hand domains, usually arranged such that two EF-hand domains may directly contact each other.

Calmodulin Target Proteins Possess a Basic Amphiphilic Helix

Circular dichroism measurements show that the conformations of EF-hand proteins change dramatically upon binding of Ca^{2+} ions. This change promotes binding of the EF-hand protein with its target protein(s). For example, calmodulin (CaM), a 148-residue protein found in many cell types, modulates the activities of a large number of target proteins, including Ca^{2+}-ATPases, protein kinases, phosphodiesterases, and NAD^+ kinase, as well as several proteins involved in intracellular motility. CaM binds to these and to many other proteins with extremely high affinities (K_D values typically in the high picomolar to low nanomolar range). All CaM target proteins possess a *basic amphiphilic alpha helix* (a **Baa helix**), to which CaM binds specifically and with high affinity. Viewed end-on, in the so-called **helical wheel** representation (Figure 37.27), a Baa helix has mostly hydrophobic residues on one face; basic residues are collected on the opposite face. However, the Baa helices of CaM target proteins, though conforming to the model, show extreme variability in sequence. How does CaM, itself a highly conserved protein, accommodate such variety of sequence and structure? Each globular domain consists of a large hydrophobic surface flanked by regions of highly negative electrostatic potential—a surface suitable for interacting with a Baa helix. The long central helix joining the two globular regions behaves as a long, flexible tether. When the target protein is bound, the two globular domains fold together, forming a single binding site for target peptides. The flexible nature of the tethering helix allows the two globular domains to adjust their orientation synergistically for maximal binding of the target protein or peptide.

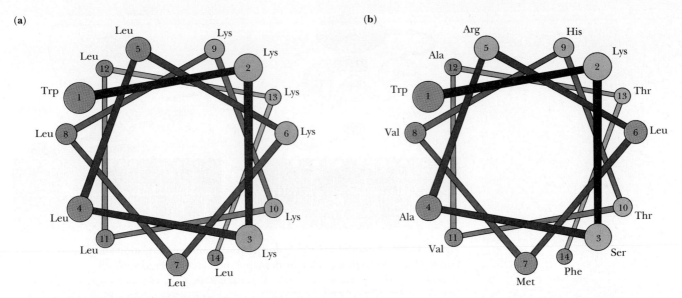

(a)

(b)

***Figure* 37.27** Helical wheel representations of (a) a model peptide, Ac-WKKLLKLLKKLLKL-CONH$_2$, and (b) the calmodulin-binding domain of spectrin. Positively charged and polar residues are indicated in green and hydrophobic residues are orange.

(Adapted from O'Neil, K., and DeGrado, W., 1990. How calmodulin binds its targets: Sequence independent recognition of amphiphilic α-helices. Trends in Biochemical Sciences 15:59–64.)

37.8 Protein Kinase C Transduces the Signals of Two Second Messengers

Protein kinase C (PKC) elicits a variety of cellular responses by phosphorylation of various target proteins at Ser and Thr residues. PKC is specifically activated by two intracellular second messengers: diacylglycerol and Ca^{2+} (the "C" in PKC stands for Ca^{2+}). Since Ca^{2+} levels increase in the cell in response to IP$_3$, the activation of PKC depends upon both of the second messengers released by hydrolysis of PIP$_2$. PKC is a cellular **transducer,** translating the hormonal message and the signals of second messengers into the protein phosphorylation events that control growth and development.

Protein kinase C (PKC) is an 80-kD polypeptide with four conserved domains and five variable regions. The conserved regions include an ATP-binding domain, a substrate-binding domain, a calcium-binding domain, and a DAG-binding domain (Figure 37.28). The DAG-binding domain is often referred to as the "pseudosubstrate domain," since it has an amino acid sequence that closely resembles protein substrates for the enzyme. With somewhat variable sequence, there are at least eleven different members of the protein kinase C family, which probably have separate and distinguishable functions. At low levels of Ca^{2+} and in the absence of DAG, protein kinase C

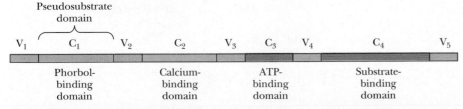

***Figure* 37.28** The primary structure of the α isozyme of protein kinase C. V indicates variable regions; C indicates conserved domains. The phorbol-binding domain is identical with the DAG-binding domain.

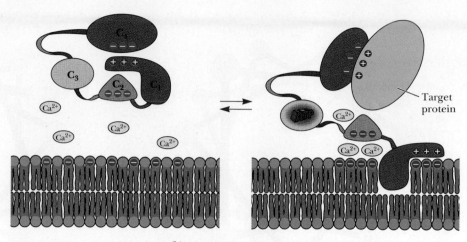

***Figure* 37.29**   A model for Ca^{2+}- and DAG-dependent activation of protein kinase C. C_1 through C_4 refer to domains indicated in Figure 37.28.

(Adapted from Sando, J. J., Maurer, M. C., Bolen, E. J., and Grisham, C. M., 1992. Role of cofactors in protein kinase C activation. Cellular Signalling 4:595–609.)

is inactive and is a soluble protein in the cytoplasm. In this state, the pseudosubstrate domain occupies the substrate-binding site, keeping the enzyme inactive (Figure 37.29). DAG binding causes conformation changes that dissociate the pseudosubstrate domain from the substrate-binding site and increase the affinity of the enzyme for Ca^{2+} and lipid, causing protein kinase C domain C_1 to bind to the cytoplasmic surface of the plasma membrane, whereupon the kinase becomes active. Control of enzymatic activity by the insertion of a pseudosubstrate domain into the active site has been referred to as **intrasteric control,** in contrast to allosteric control where an enzyme regulator with a structure unrelated to the substrate binds at a site separate from the active site. Many protein kinases and protein phosphatases are regulated by intrasteric control.

PKC phosphorylates serine and threonine residues on a wide range of protein substrates. A role for protein kinase C in cellular growth and division is demonstrated by its strong activation by **phorbol esters** (Figure 37.30). These compounds, from the seeds of *Croton tiglium,* are **tumor promoters,** agents that do not themselves cause tumorigenesis but that potentiate the effects of carcinogens. The phorbol esters mimic DAG, bind to the regulatory pseudosubstrate domain of the enzyme, and activate protein kinase C. The involvement of the protein kinase C branch of the phosphoinositide pathway in oscillatory Ca^{2+} patterns is demonstrated with phorbol esters, which themselves induce oscillatory behavior in certain cells (mouse eggs and astrocytes) but inhibit oscillations in others (liver, fibroblasts, and endothelial cells).

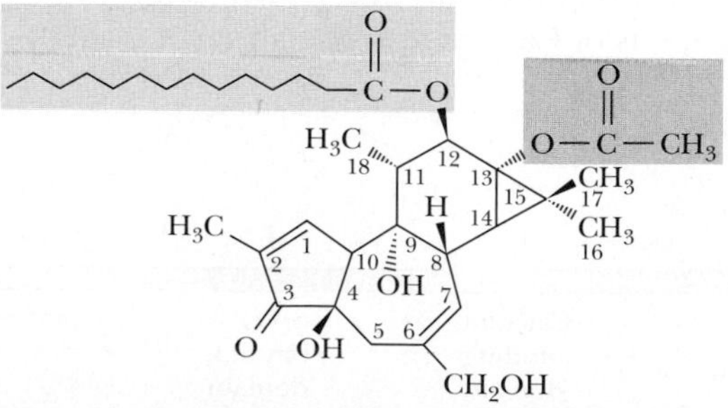

12-*O*-Tetradecanoylphorbol-13-acetate

***Figure* 37.30**   The structure of a phorbol ester. Long-chain fatty acids predominate at the 12-position, whereas acetate is usually found at the 13-position.

A Deeper Look

Okadaic Acid: A Marine Toxin and Tumor Promoter

A novel agent for evaluating the serine/threonine phosphatases is **okadaic acid** (see figure), a complex fatty acid polyketal produced by marine dinoflagellates (plankton) that accumulates in the digestive glands of shellfish and marine sponges such as *Halichondria okadaii*, from which it gets its name. Okadaic acid is the major cause of diarrhetic shellfish poisoning, and it is a potent tumor promoter. It is a specific inhibitor of PP1 and PP2A, but has no effect on other phosphatases or on protein kinases, including protein kinase C. Since PP1 and PP2A are the enzymes that reverse the actions of protein kinase C, it is no surprise that okadaic acid is as potent a tumor promoter as the phorbol esters. Okadaic acid is a hydrophobic molecule and easily enters cells, where it causes prolonged phosphorylation of many cellular proteins. In so doing, it causes long-lasting contraction of vascular smooth muscle and also mimics the stimulatory behavior of insulin on glucose metabolism. The role of okadaic acid in diarrhetic shellfish poisoning appears to be related to the diarrhetic action of cholera toxin, which activates adenylyl cyclase, and the cAMP-dependent protein kinase, which phosphorylates proteins that control sodium secretion by intestinal cells. Okadaic acid probably causes diarrhea by stimulating the phosphorylation of the same sodium channel-regulating proteins. Another toxin with similar properties is **calyculin A,** isolated from the sponge *Discodermia calyx*. It is also a strong inhibitor of PP1 and PP2 and a potent tumor promoter, although its structure differs significantly from that of okadaic acid.

Okadaic acid   : R1 = H, R2 = H
Dinophysistoxin-1 : R1 = H, R2 = CH₃
Acanthifolicin   : C9 = C10 →

Calyculin A

Structures of okadaic acid, calyculin A and related toxins.

Cellular Target Proteins Are Dephosphorylated by Phosphoprotein Phosphatases

Along with the growing appreciation of the importance of protein phosphorylation in the response of cells to hormones, growth factors, and other cellular control signals has come an equal appreciation for the roles of **phosphoprotein phosphatases.** Many phosphoprotein phosphatases, specific either for serine/threonine phosphates or for tyrosine phosphates, have now been characterized. Four different classes of serine/threonine phosphatases are known. **Type 1 phosphoprotein phosphatases (PP1)** dephosphorylate the β-subunit of phosphorylase kinase. Phosphoprotein phosphatase is inhibited by nanomolar concentrations of two heat-stable proteins, **inhibitor-1** and **inhibitor-2.** The other three classes of serine/threonine phosphatases are the **type 2 phosphatases,** and they are designated **PP2A, PP2B,** and **PP2C.** These enzymes are distinguished in part by their sensitivities to divalent cations. PP1 and PP2A do not require divalent cations, whereas PP2B is dependent on Ca^{2+} and calmodulin and PP2C requires Mg^{2+}.

The phosphatases PP2A and PP2C are predominantly cytosolic, but PP1 binds to membranes within the cell. Evidence is accumulating that PP1 (and possibly other serine/threonine phosphatases) interacts with specialized regulatory subunits that target the enzyme to particular locations in the cell and enhance its activity toward selected substrates.

37.9 The Single-TMS Receptors

Two principal classes of hormone receptors display an intrinsic enzyme activity: **receptor tyrosine kinases** and **receptor guanylyl cyclases.** Interestingly, each of these enzyme activities is manifested in two different cellular forms. Thus, guanylyl cyclase activity is found both in membrane-bound receptors and in soluble, cytoplasmic proteins. Tyrosine kinase activity, on the other hand, is exhibited by two different types of membrane proteins. The receptor tyrosine kinases are integral transmembrane proteins, whereas the nonreceptor tyrosine kinases, which are related to a family of retroviral transforming proteins, are peripheral, lipid-anchored proteins.

Nonreceptor Tyrosine Kinases

The first tyrosine kinases to be discovered were associated with **viral transforming proteins.** These proteins, produced by **oncogenic viruses,** enable the virus to *transform* animal cells, that is, to convert them to the cancerous state. A prime example is the tyrosine kinase expressed by the **src gene** of **Rous** or **avian sarcoma virus.** The protein product of this gene is **pp60^v-src** (the abbreviation refers to *p*hospho-*p*rotein, 60 kD, *v*iral origin, *s*arcoma-causing). The *src*-gene was derived from the chicken genome during the original formation of the virus. The cellular proto-oncogene homolog of pp60^v-src is referred to as pp60^c-src. pp60^v-src is a 526-residue peripheral membrane protein. It undergoes two post-translational modifications: (a) the amino group of the NH_2-terminal glycine is modified by the covalent attachment of a **myristyl** group (this modification is required for membrane association of the kinase; see Figure 37.31), and (b) Ser^17 and Tyr^416 are phosphorylated. The phosphorylation at Tyr^416, which increases kinase activity two- to threefold, appears to be an autophosphorylation. The significance of nonreceptor tyrosine kinase activity to cell growth and transformation is only partially understood, but many, many cellular proteins (of mostly undetermined function) are phosphorylated by such kinases.

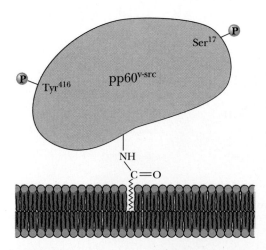

Figure 37.31 The soluble tyrosine kinase pp60^v-src is anchored to the plasma membrane via an N-terminal myristyl group.

Receptor Tyrosine Kinases

The binding of polypeptide hormones and growth factors to receptor tyrosine kinases activates the tyrosine kinase activity of these proteins. These catalytic receptors are comprised of three domains (Figure 37.32): a glycosylated extra-cellular receptor-binding domain, a transmembrane domain consisting of a single transmembrane α-helix, and an intracellular domain that includes a tyrosine kinase domain that mediates the biological response to the hormone or growth factor via its catalytic activity and a regulatory domain that contains multiple autophosphorylation sites.

There are three classes of receptor tyrosine kinases (Figure 37.32). Class I, exemplified by the **epidermal growth factor (EGF) receptor,** has an extra-cellular domain containing two Cys-rich repeat sequences. Class II, typified by the **insulin receptor,** has an $\alpha_2\beta_2$ tetrameric structure with transmembrane β-subunits and a Cys-rich domain in the extracellular α-subunit. Class III receptors, such as the **platelet-derived growth factor (PDGF) receptor,** have five (or sometimes three) immunoglobulin-like extracellular domains.

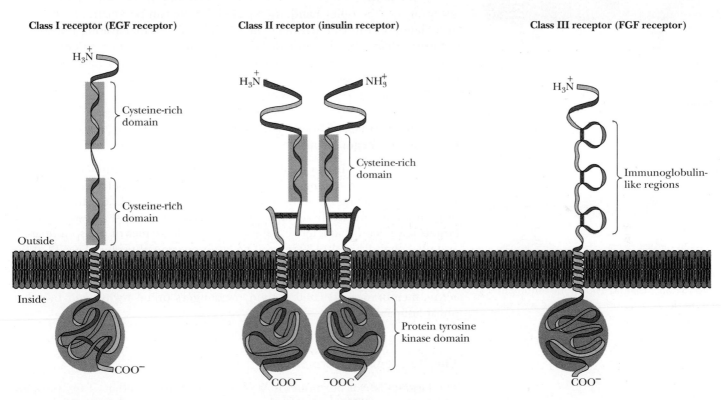

Class I receptor (EGF receptor)   **Class II receptor (insulin receptor)**   **Class III receptor (FGF receptor)**

Figure **37.32**   The three classes of receptor tyrosine kinases. Class I receptors are monomeric and contain a pair of Cys-rich repeat sequences. The insulin receptor, a typical Class II receptor, is a glycoprotein composed of two kinds of subunits in an $\alpha_2\beta_2$ tetramer. The α- and β-subunits are synthesized as a single peptide chain, together with an N-terminal signal sequence. Subsequent proteolytic processing yields the separate α- and β-subunits. The β-subunits of 620 residues each are integral transmembrane proteins, with only a single transmembrane α-helix, and with the amino terminus outside the cell and the carboxyl terminus inside. The α-subunits of 735 residues each are extracellular proteins that are linked to the β-subunits and to each other by disulfide bonds. The insulin-binding domain is located in a cysteine-rich region on the α-subunits. Class III receptors contain multiple immunoglobulin-like domains. Shown here is fibroblast growth factor (FGF) receptor, which has three immunoglobulin-like domains.

(Adapted from Ullrich, A., and Schlessinger, J., 1990. Cell 61:203–212.)

Subclass I   Ligand

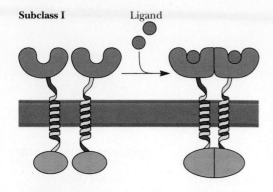

Subclass II   Ligand

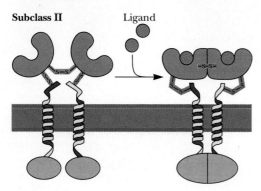

Subclass III   Ligand dimer

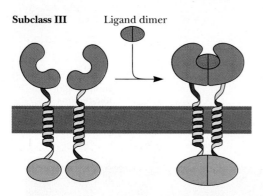

***Figure* 37.33** Ligand (hormone)-stimulated oligomeric association of receptor tyrosine kinases.

Receptor Tyrosine Kinases Are Membrane-Associated Allosteric Enzymes

Given that the extracellular and intracellular domains of receptor tyrosine kinases are joined only by a single transmembrane helical segment, how does extracellular hormone binding activate intracellular tyrosine kinase activity? How is the signal transduced? As shown in Figure 37.33, signal transduction occurs by hormone-induced oligomeric association of receptors. Binding of hormone triggers a conformational change in the *extracellular* domain, which induces oligomeric association. Oligomeric association allows adjacent *cytoplasmic* domains to interact, leading to phosphorylation of the cytoplasmic domains, and stimulation of cytoplasmic tyrosine kinase activity. In the case of the class II receptors (for example, the insulin receptor), hormone binding induces interactions between the two $\alpha\beta$ halves within the disulfide-linked receptor complex. By virtue of these ligand-induced conformation changes and oligomeric interactions, receptor tyrosine kinases are **membrane-associated allosteric enzymes.**

Autophosphorylation at Tyr residues allows the receptor tyrosine kinase to remain active even after the activating hormone has dissociated from the receptor. On the other hand, the tyrosine kinase can be inactivated by phosphorylation of intracellular Ser/Thr residues. These inactivating phosphorylations are catalyzed by protein kinase C and by cAMP-dependent protein kinases, providing a direct linkage between receptor tyrosine kinases and several key second messengers, including cAMP, IP_3, DAG, and Ca^{2+}.

Receptor Tyrosine Kinases Phosphorylate a Variety of Cellular Target Proteins

Receptor tyrosine kinases catalyze the phosphorylation of numerous cellular target proteins, producing coordinated changes in cell behavior, including alterations in membrane transport of ions and amino acids, the transcription of genes, and the synthesis of proteins. Several individual phosphorylation targets have been characterized, including the γ-isozymes of phospholipase C and phosphatidylinositol-3-kinase. The latter enzyme phosphorylates PI at the 3-position, producing several phosphoinositide metabolites that are not hydrolyzed by known phospholipase C enzymes. These PI-3-P species may either act as membrane-associated second messengers or be hydrolyzed by other phospholipases.

The Polypeptide Hormones

The largest class of hormones in vertebrate organisms is that of the **polypeptide hormones** (Table 37.4). One of the first polypeptide hormones to be discovered, **insulin,** was described by Banting and Best in 1921. Insulin, a secretion of the pancreas, controls glucose utilization and promotes the synthesis of proteins, fatty acids, and glycogen. Insulin, which is typical of the **secreted polypeptide hormones,** was discussed in detail in Chapters 4 and 12.

Many other polypeptide hormones are produced and processed in a manner similar to that of insulin. Three unifying features of their synthesis and cellular processing should be noted. First, all secreted polypeptide hormones are originally synthesized with a signal sequence, which facilitates their eventual direction to secretory granules, and thence to the extracellular milieu. Second, peptide hormones are usually synthesized from mRNA as inactive precursors, termed **preprohormones,** which become activated by proteolysis. Third, a single polypeptide precursor or preprohormone may produce sev-

Table 37.4
Polypeptide Hormones

Hormone	Amino Acid Residues	Source	Target Cells	Function
Adrenocorticotropic hormone (ACTH)	39	Anterior pituitary	Adrenal cortex	Promotes adrenal steroid production
Angiotensin II	8	Sequence of conversions in kidney, blood, and lung	Adrenal cortex	Stimulates aldosterone production
Atrial natriuretic factor (ANF)	28	Atrial walls of heart	Primarily kidney	Regulates Na^+ excretion
Bradykinin	9	Kidney, other tissues	Blood vessels	Causes vasodilation
Calcitonin	33	Thyroid gland	Bone	Regulates plasma Ca^{2+} and phosphate
Chorionic gonadotropin	α, 96 β, 147	Placenta	Various reproductive tissues	Maintains pregnancy
Follicle-stimulating hormone (FSH)	α, 96 β, 120	Anterior pituitary	Gonads	Stimulates growth and development
Gastrin	17	Gastrointestinal tract	GI tract, gallbladder, pancreas	Regulates digestion
Glucagon	29	Pancreas	Primarily liver	Regulates metabolism and blood glucose
Growth hormone (GH)	191	Anterior pituitary	Many: bone, fat, liver	Stimulates skeleton and muscle growth
Insulin	A, 21 B, 30	Pancreas	Primarily liver, muscle, and fat	Regulates metabolism and blood glucose
Luteinizing hormone (LH)	α, 96 β, 121	Anterior pituitary	Gonads, ovarian follicle cells	Triggers ovulation
Oxytocin	9	Synthesized in hypothalamus; secreted from posterior pituitary	Breast, uterus	Triggers "let-down" of milk; stimulates uterine contraction
Parathyroid hormone (PTH)	84	Parathyroid glands	Bone, kidney	Regulates plasma Ca^{2+} and phosphate
Prolactin	197	Anterior pituitary	Breast	Stimulates milk production
Somatostatin	14	Hypothalamus	Anterior pituitary	Inhibits growth hormone secretion
Thyroid-stimulating hormone (TSH)	α, 96 β, 112	Anterior pituitary	Thyroid gland	Promotes thyroid hormone production
Vasopressin	9	Synthesized in hypothalamus; secreted from posterior pituitary	Kidney	Increases water reabsorption

Source: Adapted from Rhoades, R., and Pflanzer, R., 1992. *Human Physiology*, 2nd ed. Philadelphia: Saunders College Publishing.

eral different peptide hormones by suitable proteolytic processing. The following processing events are common to all preprohormones:

1. Proteolytic cleavage of a hydrophobic N-terminal signal peptide sequence
2. Proteolytic cleavage at a site defined by pairs of basic amino acid residues
3. Proteolytic cleavage at the site of single Arg residues

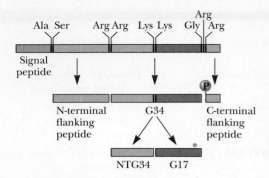

Figure 37.34 The major pathway of gastrin biosynthesis in human gastric mucosal cells. The asterisk (*) indicates a sulfate group, and P indicates a phosphorylation site.

(Adapted from Dockray, G., et al., 1989. Gastrin and CCK-related peptides. In Martinez, J., ed. Peptide Hormones as Prohormones. Halstead Press.)

4. Post-translational modification of individual amino acids, including α-amidation of the C-terminal residue, phosphorylation, glycosylation, or acetylation of the N-terminal residue

The Processing of Gastrin

A good example of these processing events is the biosynthesis of **gastrin,** a heptadecapeptide (17 residues) secreted by the antral mucosa in the stomach. Gastrin activates acid secretion in the stomach, and also stimulates growth of the acid-secreting mucosa. It is a product of **preprogastrin,** a peptide of 101 (man) or 104 (pig) residues (Figure 37.34). As expected for a secretory hormone, it has a signal peptide (21 residues), cleavage of which leaves **progastrin.** Cleavage at pairs of basic residues (noted in the figure) followed by C-terminal amidation produces gastrin (G17) and three other peptides. Several post-translational modifications occur in progastrin, including incorporation of a sulfate at Tyr^{12}, phosphorylation of Ser^{21}, and protective modifications of both the C- and N-termini of gastrin itself. The N-terminal residue of gastrin is a **pyroglutamate** (a pyrrolidone carboxylic acid), formed by cyclization of the N-terminal glutamine after proteolytic cleavage of the flanking peptide. An identical N-terminal modification is found in bacteriorhodopsin (Chapter 9). At the C-terminal end, three enzymatic steps result in amidation, which protects the peptide hormone from action of carboxypeptidases. The steps involve a trypsin-like cleavage on the C-terminal side of the second arginine, a carboxypeptidase-like action to cleave the two arginines, and degradation of the glycine residue to leave a C-terminal phenylalanine amide (Figure 37.35).

An impressive example of the production of many hormone products from a single precursor is the case of **prepro-opiomelanocortin,** a 250-residue precursor peptide synthesized in the pituitary gland. A cascade of proteolytic steps produces, as the name implies, a natural *opi*ate substance **(endorphin),** *melano*cyte-stimulating hormones (α- and β-MSH), and *corti*cotropin (also known as *a*dreno*corti*cotrophic *h*ormone, *ACTH*) as shown in Figure 37.36. As noted, these proteolytic cleavages actually take place in different tissues. Cleavage of pro-opiomelanocortin in the anterior pituitary yields corticotropin and β-lipotropin, which proceed to cells of the central nervous system for the final proteolytic steps.

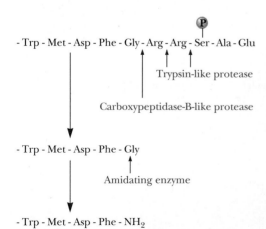

Figure 37.35 The enzyme-catalyzed proteolysis and amidation reactions at the C-terminus of progastrin that lead to gastrin.

Protein-Tyrosine Phosphatases

Phosphatases specific for phosphorylated tyrosines on proteins are different from their serine/threonine-specific cousins, and display a wide variety of structures. Some of these **protein-tyrosine phosphatases (PTPases)** are integral membrane proteins, whereas others are cytoplasmic in nature (Figure 37.37 and Table 37.5). The cytoplasmic PTPases consist of N-terminal catalytic domains with C-terminal regulatory regions that may guide the enzyme to specific intracellular locations and substrates and/or specifically activate the PTPase toward those substrates. The membrane-bound PTPases all consist of a cytoplasmic catalytic domain, a transmembrane domain, and various extracellular domains that are designed to convey signals across the membrane. The extracellular domains bear considerable similarity to **cell adhesion molecules (CAM),** with repeated immunoglobulin domains and fibronectin type III repeats. PTPases have been linked to the inhibition of cell growth by cell–cell contact and to regulation of T-cell activation and proliferation. It has been postulated that membrane-bound PTPases may act as tumor suppressors, and that their loss or mutation could lead to unrestrained cell proliferation and transformation.

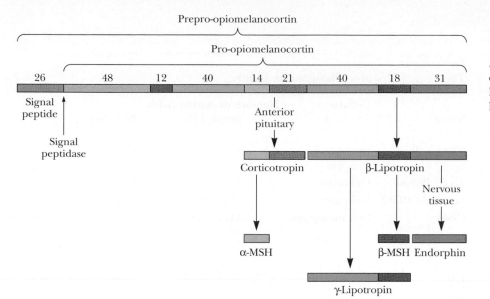

Figure 37.36 The conversion of prepro-opiomelanocortin to a family of peptide hormones, including corticotropin, β- and γ-lipotropin, α- and β-MSH, and endorphin.

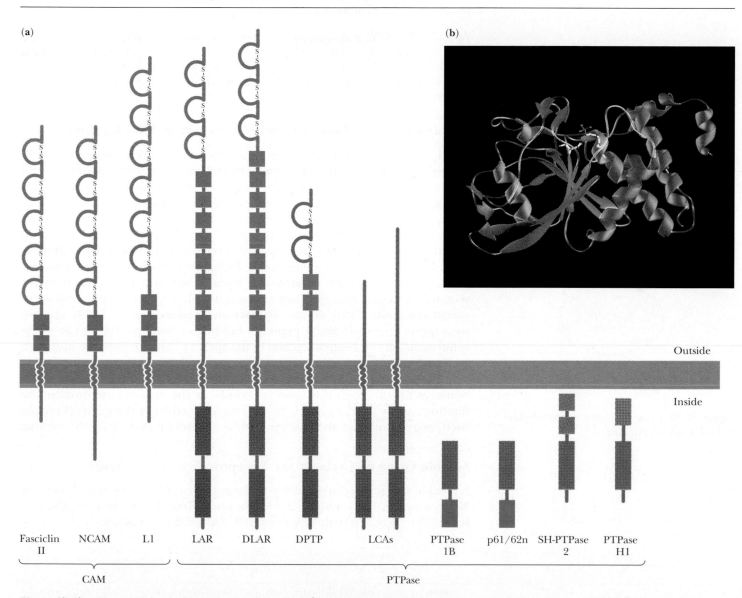

(a)

(b)

Fasciclin II · NCAM · L1 · LAR · DLAR · DPTP · LCAs · PTPase 1B · p61/62n · SH-PTPase 2 · PTPase H1

CAM

PTPase

Figure 37.37 The primary structures of cytoplasmic and membrane-bound protein-tyrosine phosphatases (PTPases) and cell adhesion molecules (CAM). Key: S-S loops are immunoglobulin domains; blue squares are fibronectin domains; gold rectangles are PTPase catalytic domains; green and red rectangles are membrane-associated domains; blue squares on SH-PTPase are SH2 domains; olive rectangle on PTPase H1 is a cytoskeletal domain.

(Adapted from Streuli, M., et al., 1989. A family of receptor-linked protein tyrosine phosphatases in humans and Drosophila. Proceedings of the National Academy of Sciences, U.S.A. 86:8698–8702; and Mauro, L. J., and Dixon, J. E., 1994. Zip codes direct intracellular protein tyrosine phosphatases to the correct cellular "address." Trends in Biochemical Sciences 19:151–155. Molecular graphic courtesy of David Barford, Oxford University.)

Table 37.5

The Protein-Tyrosine Phosphatase Family

Name	Cellular Location	Number of Amino Acids (mass, kD)	Function
PTPase IB	Cytosolic face of ER	432 (50)	Unknown
T cell PTPase	Unknown	415 (48.4)	Unknown
Rat brain PTP-1	Unknown	432 (50)	Unknown
CD45	Cell membrane	1120–1281 (180–200)	Mediation of B and T lymphocyte interactions? T cell activation?
LAR	Cell membrane	1881 (variable)	Inhibition of growth by cell–cell contact?
DLAR	Cell membrane	1997 (variable)	Inhibition of growth by cell–cell contact?
DPTP	Cell membrane	1439 (variable)	Inhibition of growth by cell–cell contact?

Adapted from Banks, P., 1990. Tyrosine pyrophosphates: cellular superstars in the offing. *Journal of NIH Research* **2**:62–66.

Membrane-Bound Guanylyl Cyclases Are Single-TMS Receptors

Another cellular second messenger, **guanosine 3′,5′-cyclic monophosphate (cGMP)**, is formed from GTP by **guanylyl cyclase,** an enzyme found in several different forms in different cellular locations. **Membrane-bound guanylyl cyclases** constitute a second class of single-TMS receptors with an extracellular hormone-binding domain; a single, α-helical transmembrane segment; and an intracellular catalytic domain (Figure 37.38). A variety of peptides act to stimulate the membrane-bound guanylyl cyclases, including **atrial natriuretic peptide (ANP),** which regulates body fluid homeostasis and cardiovascular function; the **heat-stable enterotoxins** from *E. coli;* and a series of peptides secreted by mammalian ova (eggs), which stimulate sperm motility and act as sperm chemoattractant signals. **Speract** and **resact** (Figure 37.39) are two such sperm chemoattractant peptides. Binding of these peptides to an extracellular site on the guanylyl cyclase in the sperm plasma membrane induces a conformational change that activates the intracellular catalytic site for cyclase activity. Activation may involve oligomerization of receptors in the membrane, as for the RTKs discussed previously. In the case of enterotoxins, the binding activity for the peptide has been separated from the guanylyl cyclase itself, suggesting that the toxin receptor is distinct from guanylyl cyclase.

Soluble Guanylyl Cyclases Are Receptors for Nitric Oxide

A soluble guanylyl cyclase in the cytoplasm is the receptor for **nitric oxide,** or **NO ·,** a reactive free radical that plays two different roles in cells. On one hand, NO · acts as a neurotransmitter (Chapter 38) and as a second messen-

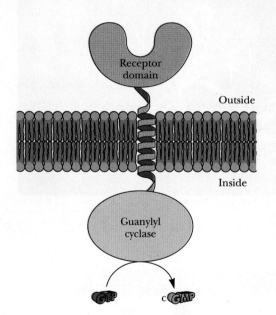

Figure **37.38** The structure of membrane-bound guanylyl cyclases.

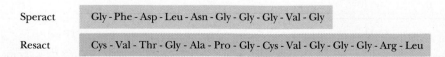

Speract	Gly - Phe - Asp - Leu - Asn - Gly - Gly - Gly - Val - Gly
Resact	Cys - Val - Thr - Gly - Ala - Pro - Gly - Cys - Val - Gly - Gly - Gly - Arg - Leu

Figure **37.39** The primary structures of two guanylyl cyclase-activating peptides. Their receptors are glycoproteins with masses of 120 kD to 180 kD.

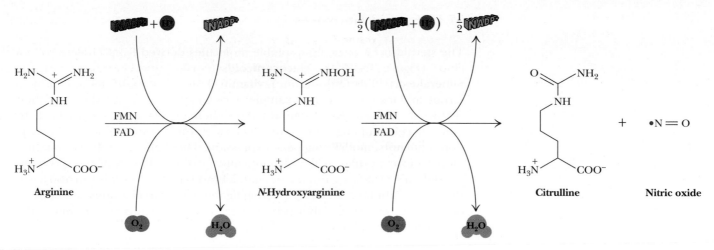

Figure 37.40 The synthesis of nitric oxide (NO·) by NO synthase.

ger, inducing relaxation of vascular smooth muscle and mediating penile erection. On the other hand, NO· enables large white blood cells known as macrophages to kill tumor cells and bacteria.

Cyclic GMP generated by the NO-stimulated, soluble guanylyl cyclase is itself a second messenger that can activate or inhibit a variety of processes. For example, cGMP can (a) regulate ion-channel gating in cerebellar glial cells, and (b) block gap junction conductivity in retinal cells.

NO· is synthesized from arginine by **NO synthase** in two consecutive monooxygenase reactions (Figure 37.40). As a dissolved gas, NO· is capable of rapid diffusion across membranes in the absence of any apparent carrier mechanism. This property makes NO· a particularly attractive second messenger, since NO· generated in one cell can exert its effects quickly in many neighboring cells. NO· has a very short cellular half-life (1 to 5 seconds) and is rapidly degraded by nonenzymatic pathways.

Binding of NO· to the heme prosthetic group of soluble guanylyl cyclase causes a fiftyfold increase in the rate of cGMP synthesis. Soluble guanylyl cyclase is a 150-kD dimer consisting of an α-subunit (82 kD), a β-subunit (70 kD), and a heme prosthetic group. NO· binds to the heme group to form a nitrosoheme (Figure 37.41). Sequence analysis indicates that the carboxyl region of the 70-kD subunit is homologous with the carboxyl domain of membrane-bound forms of guanylyl cyclase.

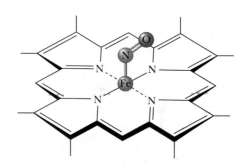

Figure 37.41 Binding of NO· to the heme moiety of guanylyl cyclase to form a nitrosoheme.

A Deeper Look

Nitric Oxide, Nitroglycerin, and Alfred Nobel

NO· is the active agent released by **nitroglycerin** (see figure), a powerful drug that ameliorates the symptoms of heart attacks and **angina pectoris** (chest pain due to coronary artery disease) by causing the dilation of coronary arter-

ies. Nitroglycerin is also the active agent in dynamite. Ironically, Alfred Nobel, the inventor of dynamite who also endowed the Nobel Prizes, himself suffered from angina pectoris. In a letter to a friend in 1885, Nobel wrote, "It sounds like the irony of fate that I should be ordered by my doctor to take nitroglycerin internally."

The structure of nitroglycerin, a potent vasodilator.

37.10 Steroid Hormones

The steroid hormones, lipid-soluble molecules derived from cholesterol, include (Figure 37.42) the **glucocorticoids** (cortisol and corticosterone), the **mineralocorticoids** (aldosterone), **vitamin D,** and the **sex hormones** (progesterone and testosterone, for example) (see Chapter 24 for the details of their synthesis). The steroid hormones exert their effects in two ways. First, by entering cells and migrating to the nucleus, steroid hormones act as transcription regulators, modulating gene expression. These effects of the steroid hormones occur on time scales of hours and involve synthesis of new proteins. Considerable evidence has accumulated, however, that steroids can also act at the cell membrane, directly regulating ligand-gated ion channels and perhaps other processes. These latter processes take place very rapidly, on time scales of seconds and minutes.

Receptor Proteins Carry Steroids to the Nucleus

Intracellular effects of the steroid hormones are initiated when the steroid diffuses across the plasma membrane and binds to specific receptor proteins. The binding of steroids to these receptors is typically very tight, with dissocia-

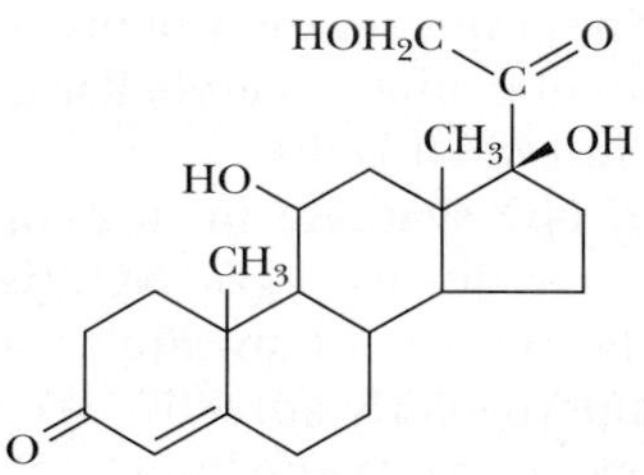

Cortisol

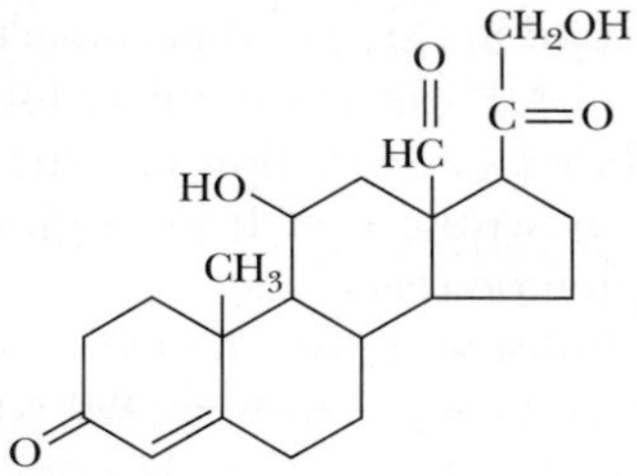

Aldosterone

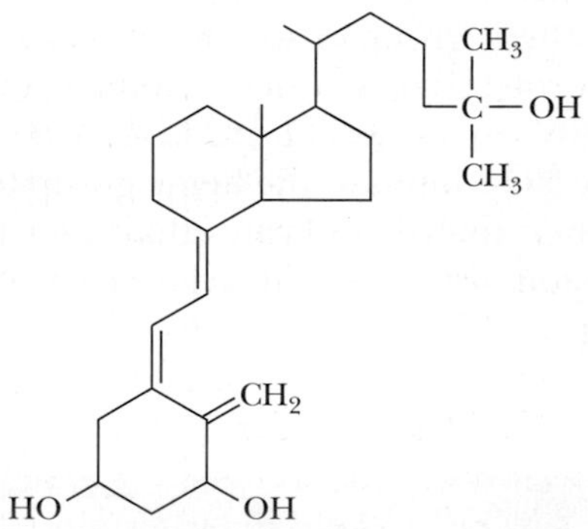

1,25-Dihydroxyvitamin D$_3$

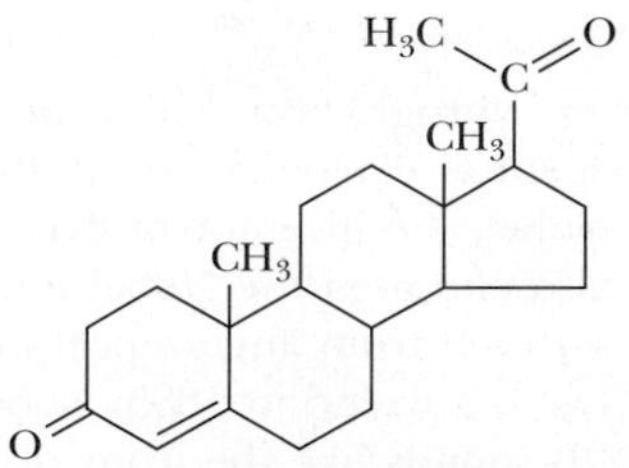

Progesterone

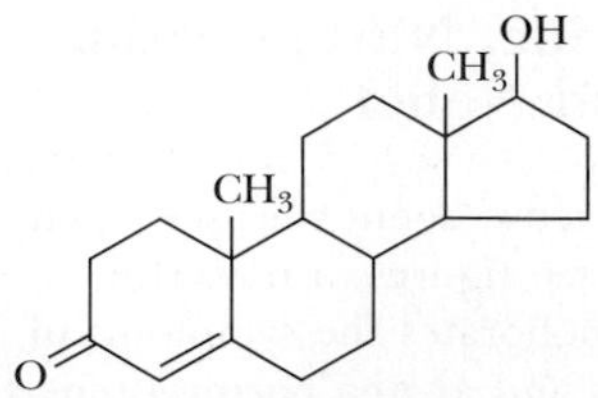

Testosterone

***Figure* 37.42** Structures of some steroid hormones.

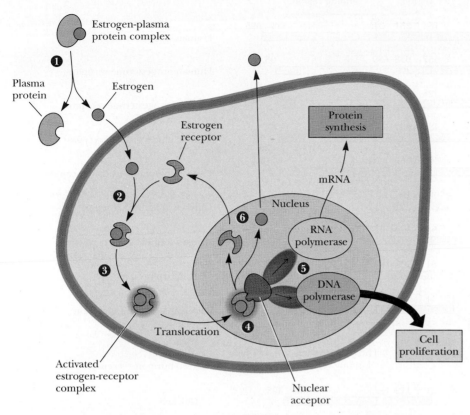

***Figure* 37.43**   A model for steroid hormone action in target cells. The hormone (for example, estrogen) dissociates from plasma membrane proteins, diffuses into the cell, and binds to receptor proteins. The active hormone:receptor complex migrates to the nucleus where it interacts with DNA or transcription factors or both.

(Adapted from Welshons, W., and Jordan, V., 1987. Heterogeneity of nuclear steroid hormone receptors with an emphasis on unfilled receptor sites, In Clark, C., ed. Steroid Hormone Receptors. Ellis Horwood, VCH Publishers.)

tion constants in the nanomolar range. Nearly all the receptor molecules in a steroid-sensitive cell are located in the nucleus. Nonetheless, because of the highly hydrophobic character of the steroids themselves and the low likelihood that they could migrate through the cytoplasm to the nucleus without the help of receptor proteins, it is believed that small concentrations of receptor proteins are available in the cytoplasm to ferry the steroids from the plasma membrane to the nucleus (Figure 37.43).

Steroid receptor proteins possess a strikingly similar structural organization (Figure 37.44), indicating that they are all members of a *gene superfamily,* and that they have evolved from a common ancestral precursor. Each of these receptor proteins contains a hydrophobic domain near the C-terminus that is presumed to bind the steroid hormone, as well as a central, hydrophilic DNA-binding domain. These DNA-binding domains are highly homologous (Figure 37.45). Perhaps the most striking feature of the DNA domains is the conserved arrangement of nine Cys residues in all steroid receptors. Three pairs of these Cys residues are in Cys-X-X-Cys sequences, which are commonly found in the DNA-binding **zinc-finger proteins** (Chapter 31). The steroid-binding domains of the steroid receptor proteins show less sequence homology, but certain residues are conserved at equivalent positions in all receptors. These domains are presumed to form hydrophobic pockets specific to each steroid hormone.

The steroid-receptor complex has two functions in the nucleus. It can bind directly to DNA in order to regulate transcription, or it may combine

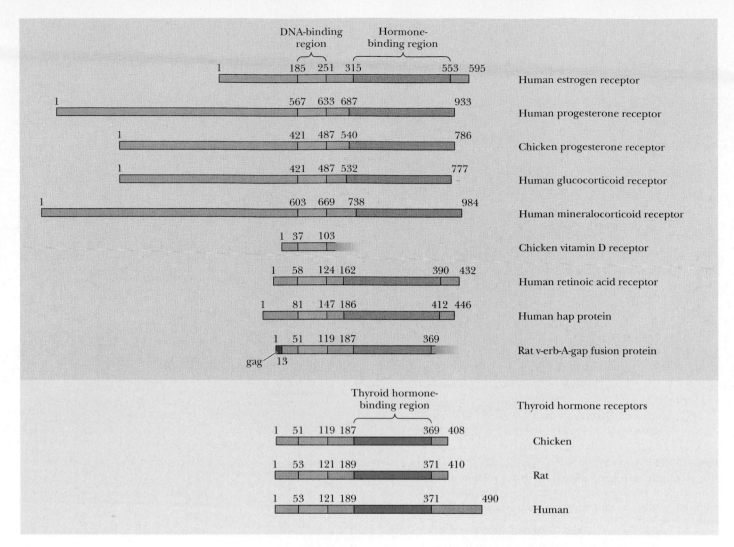

Figure 37.44 The primary structures of nuclear steroid and thyroid receptor proteins.

(Figures 37.44 and 37.45 adapted from Gronemeyer, H., ed., 1988. Affinity labelling and cloning of steroid and thyroid hormone receptors. Ellis Horwood, VCH Publishers.)

Region C

Human estrogen receptor	185	C A V C N D Y A S G Y H Y G V W S C E G C K A F F K R S I Q G – – H N D Y M C P A T N Q C T I D K N R R K S C Q A C R L R K C Y E V G M
Chicken progesterone receptor	581	C L I C G D E A S G C H Y G V L T C G S C K V F F K R A M E G Q – H N – Y L C A G R N D C I V D K I R R K N C P A C R L R K C C Q A G M
Human glucocorticoid receptor	421	C L V C S D E A S G C H Y G V L T C G S C K V F F K R A V E G Q – H N N Y L C A G R N D C I I D K I R R K N C P A C R Y R K C L Q A G M
v-erb-A protein	37	C V V C G D K A T G Y H Y R C I T C E G C K S F F R R T I Q K N L H P T Y S C T Y D G C C V I D K I T R N Q C Q L C R F K K C I S V G M

Figure 37.45 Primary structure alignment of the DNA-binding domains of nuclear receptor proteins.

Figure 37.46 The structure of triiodothyronine and thyroxine.

R = H   Triiodothyronine (T$_3$)
R = I   Thyroxine (T$_4$)

with **transcription factors** such as the *Jun* and *Fos* proteins (Table A in box, page 1190). In these latter interactions, the steroid hormone receptors regulate gene expression without interacting directly with DNA.

The receptor proteins for **thyroid hormones** are highly homologous with those of the steroid hormones (Figure 37.44). These thyroid hormone receptors, which mediate the effects of **thyroxine (T$_4$)** and **triiodothyronine (T$_3$)** (Figure 37.46), possess DNA-binding domains with the same Cys-X-X-Cys motif and identically conserved Lys and Arg residues. In spite of the high degree of homology in their DNA-binding domains, these receptor proteins specifically *recognize unique DNA sequences on target genes* (Chapter 31).

A Deeper Look

The Acrosome Reaction

Steroid hormones affect ion channels in the **acrosome reaction,** which must occur before human sperm can fertilize an egg. The **acrosome** is an organelle that surrounds the head of a sperm (see figure) and lies just inside and juxtaposed with the plasma membrane. The acrosome itself is essentially a large vesicle of hydrolytic and proteolytic en-

zymes. In the acrosome reaction, influx of Ca^{2+} ions causes the outer acrosomal membrane to fuse with the plasma membrane. These fused membrane segments separate from each other and diffuse away, freeing the acrosomal enzymes to attack the egg, and exposing binding sites on the inner acrosomal membrane that are thought to interact with the egg in the fertilization process.

This acrosome reaction is induced

by **progesterone,** a female hormone secreted by the cumulus oophorus, a collection of ovarian follicle cells surrounding the egg! Intracellular Ca^{2+} levels increase within seconds of treating human sperm with progesterone. These effects must occur via binding of the steroid to the sperm plasma membrane. A far longer time would be required for progesterone to act through an enhancement of transcription.

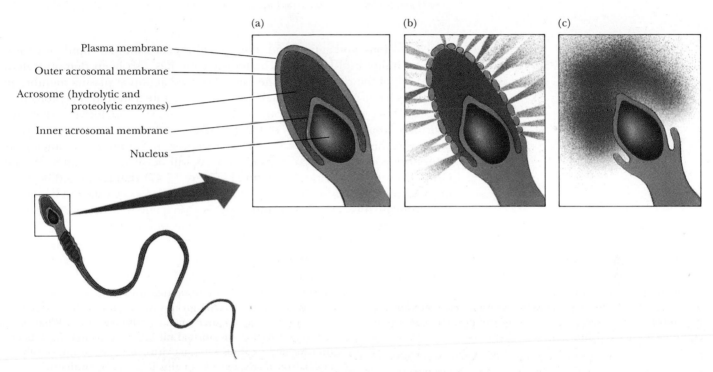

Plasma membrane

Outer acrosomal membrane

Acrosome (hydrolytic and proteolytic enzymes)

Inner acrosomal membrane

Nucleus

The acrosome reaction in human sperm.

(Adapted from Meizel, S., 1985. Molecules that initiate or help stimulate the acrosome reaction by their interaction with the mammalian sperm surface. American Journal of Anatomy 174:285–302.)

Extracellular Effects of the Steroid Hormones

Several effects of steroid hormones appear to involve action at the plasma membrane. For example, progesterone modulates Ca^{2+} channels in the membranes of brain stem neurons and also activates processes in *Xenopus laevis* oocytes by binding to the plasma membrane. Certain steroid actions occur so rapidly that activation of protein synthesis cannot be involved. The male steroid hormone testosterone quickly stimulates the transport of glu-

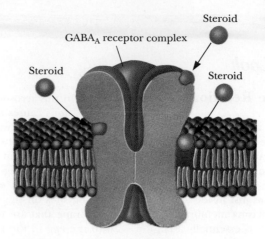

Figure 37.47 A schematic model of the GABA$_A$ receptor, indicating the interactions of steroid hormones with the receptor.

(Adapted from Touchette, N., 1990. Man bites dogma; a new role for steroid hormones. Journal of NIH Research 2:71–74.)

cose, Ca^{2+} ions, and amino acids into rat kidney cells. Similar rapid induction of Ca^{2+} influx into heart cells by testosterone has also been demonstrated.

Much of the resistance to the idea of steroid action at plasma membrane receptors has arisen from an inability to demonstrate a steroid-receptor interaction. However, several groups have now shown conclusively that **3α-hydroxy, 5α-pregnan-20-one (3α-OH-DHP),** a metabolite of progesterone, binds with high affinity to the GABA$_A$ receptor in the brain, enhancing its inhibitory effect on neural transmission. As will be seen in Chapter 38, the GABA$_A$ receptor is a chloride channel (Figure 37.47) that is opened by binding **γ-aminobutyric acid** (GABA). The demonstration that a steroid hormone can stimulate this channel is a landmark discovery, and may presage a revision of the previous dogma that steroids act only at intracellular sites.

Problems

1. Compare and contrast the features and physiological advantages of each of the major classes of hormones, including the steroid hormones, polypeptide hormones, and the amino acid–derived hormones.

2. Compare and contrast the features and physiological advantages of each of the known classes of second messengers.

3. Nitric oxide may be merely the first of a new class of gaseous second messenger/neurotransmitter molecules. Based on your knowledge of the molecular action of nitric oxide, suggest another gaseous molecule that might act as a second messenger and propose a molecular function for it.

4. Herbimycin A is an antibiotic that inhibits tyrosine kinase activity by binding to SH groups of cysteine in the *src* gene tyrosine kinase and other similar tyrosine kinases. What effect might it have on normal rat kidney cells that have been transformed by Rous sarcoma virus? Can you think of other effects you might expect for this interesting antibiotic?

5. Monoclonal antibodies that recognize phosphotyrosine are commercially available. How could such an antibody be used in studies of cell signaling pathways and mechanisms?

6. Explain and comment on this statement: The main function of hormone receptors is that of signal amplification.

Further Reading

Asaoka, Y., et al., 1992. Protein kinase C, calcium and phospholipid degradation. *Trends in Biochemical Sciences* **17**:414–417.

Bell, R., and Burns, D., 1991. Lipid activation of protein kinase C. *Journal of Biological Chemistry* **266**:4661–4664.

Bourne, H., Sanders, D., and McCormick, F., 1991. The GTPase superfamily: Conserved structure and molecular mechanism. *Nature* **349**:117–127.

Cadena, D., and Gill, G., 1992. Receptor tyrosine kinases. *The FASEB Journal* **6**:2332–2337.

Cohen, P., 1992. Signal integration at the level of protein kinases, protein phosphatases and their substrates. *Trends in Biochemical Sciences* **17**:408–413.

Cohen, P., and Cohen, P. T. W., 1989. Protein phosphatases come of age. *Journal of Biological Chemistry* **264**:21435–21438.

DeVos, A., Ultsch, M., and Kossiakoff, A., 1992. Human growth hormone and extracellular domain of its receptor: Crystal structure of the complex. *Science* **255**:306–312.

Eichmann, K., 1993. Transmembrane signaling of T lymphocytes by ligand-induced receptor complex assembly. *Angewandte Chemie, International Edition* **32**:54–63.

Exton, J., 1990. Signaling through phosphatidylcholine breakdown. *Journal of Biological Chemistry* **265**:1–4.

Galione, A., 1993. Cyclic ADP-ribose: A new way to control calcium. *Science* **259**:325–326.

Garbers, D., 1989. Guanylate cyclase, a cell surface receptor. *Journal of Biological Chemistry* **264**:9103–9106.

Gilman, A., 1987. G proteins: Transducers of receptor-generated signals. *Annual Review of Biochemistry* **56**:615–649.

Hakamori, S., 1990. Bifunctional role of glycosphingolipids. *Journal of Biological Chemistry* **265**:18713–18716.

Hepler, J., and Gilman, A., 1992. G proteins. *Trends in Biochemical Sciences* **17**:383–387.

Hollenberg, M., 1991. Structure-activity relationships for transmembrane signaling: The receptor's turn. *The FASEB Journal* **5**:178–186.

Hunter, T., and Cooper, J., 1985. Protein-tyrosine kinases. *Annual Review of Biochemistry* **54**:897–930.

Kaziro, Y., et al., 1991. Structure and function of signal-transducing GTP-binding proteins. *Annual Review of Biochemistry* **60**:349–400.

Kemp, B., and Pearson, R., 1991. Intrasteric regulation of protein kinases and phosphatases. *Biochimica et Biophysica Acta* **1094**:67–76.

Kennelly, P., and Krebs, E., 1991. Consensus sequences as substrate specificity determinants for protein kinases and protein phosphatases. *Journal of Biological Chemistry* **266**:15555–15558.

Knowles, R., and Moncada, S., 1992. Nitric oxide as a signal in blood vessels. *Trends in Biochemical Sciences* **17**:399–402.

Koch, C., et al., 1991. SH2 and SH3 domains: Elements that control interactions of cytoplasmic signaling proteins. *Science* **252**:668–674.

Linder, M., and Gilman, A., 1992. G proteins. *Scientific American* **267**:56–65.

Liskovitch, M., 1992. Crosstalk among multiple signal-activated phospholipases. *Trends in Biochemical Sciences* **17**:393–399.

Marshall, M. S., 1993. The effector interactions of p21[ras]. *Trends in Biochemical Sciences* **18**:250–254.

Mauro, L. J., and Dixon, J. E., 1994. Zip codes direct intracellular protein tyrosine phosphatases to the correct "address." *Trends in Biochemical Sciences* **19**:151–155.

Mathias, S., et al., 1993. Activation of the sphingomyelin signaling pathway in intact EL4 cells and in a cell-free system by IL-1β. *Science* **259**:519–522.

Menniti, F., et al., 1993. Inositol phosphates and cell signaling: New views of InsP$_5$ and InsP$_6$. *Trends in Biochemical Sciences* **18**:53–56.

Mustelin, T., and Burn, P., 1993. Regulation of *Src* family tyrosine kinases in lymphocytes. *Trends in Biochemical Sciences* **18**:215–220.

Pazin, M., and Williams, L., 1992. Triggering signaling cascades by receptor tyrosine kinases. *Trends in Biochemical Sciences* **17**:374–378.

Rhee, S., and Choi, K., 1992. Regulation of inositol phospholipid-specific phospholipase C isozymes. *Journal of Biological Chemistry* **267**:12393–12396.

Roach, P., 1991. Multisite and hierarchical protein phosphorylation. *Journal of Biological Chemistry* **266**:14139–14142.

Samelson, L., and Klausner, R., 1992. Tyrosine kinases and tyrosine-based activation motifs. *Journal of Biological Chemistry* **267**:24913–24916.

Satoh, T., Nakafuku, M., and Kaziro, Y., 1992. Function of *ras* as a molecular switch in signal transduction. *Journal of Biological Chemistry* **267**:24149–24152.

Schlessinger, J., 1993. How receptor tyrosine kinases activate Ras. *Trends in Biochemical Sciences* **18**:273–275.

Schlichting, I., et al., 1990. Time-resolved X-ray crystallographic study of the conformational change in Ha-*Ras* p21 protein on GTP hydrolysis. *Nature* **345**:309–314.

Stamler, J., Singel, D., and Loscalzo, J., 1992. Biochemistry of nitric oxide and its redox-active forms. *Science* **258**:1898–1902.

Sternweiss, P. C., and Smrcka, A. V., 1992. Regulation of phospholipase C by G proteins. *Trends in Biochemical Sciences* **17**:502–506.

Takasawa, S., et al., 1993. Cyclic ADP-ribose in insulin secretion from pancreatic β cells. *Science* **259**:370–373.

Taylor, C., and Marshall, I., 1992. Calcium and inositol 1,4,5-trisphosphate receptors: A complex relationship. *Trends in Biochemical Sciences* **17**:403–407.

Traylor, T., and Sharma, V., 1992. Why NO? *Biochemistry* **31**:2847–2849.

Ullrich, A., and Schlessinger, J., 1990. Signal transduction by receptors with tyrosine kinase activity. *Cell* **61**:203–212.

Wittinghofer, A., and Pai, E., 1991. The structure of *ras* protein: A model for a universal molecular switch. *Trends in Biochemical Sciences* **16**:382–387.

Yarden, Y., and Ullrich, A., 1988. Growth factor receptor tyrosine kinases. *Annual Review of Biochemistry* **57**:443–478.

Chapter 38

Excitable Membranes, Neurotransmission, and Sensory Systems

• •

Spiders and Snakes from "Locupletissimi Rerum Naturalium" by Albert Seba (c. 1750)

Outline

T he survival of higher organisms is predicated on the ability to respond *rapidly* to sensory input such as sights, sounds, and smells. The responses to such stimuli may include muscle movements and many forms of intercellular communication. Hormones (as described in the previous chapter) can move through an organism only at speeds determined by the circulatory system. In most higher organisms, a *faster* means of communication is crucial. Nerve impulses, which can be propagated at speeds up to 100 m/sec, provide a means of intercellular signaling that is fast enough to encompass sensory recognition, movement, and other physiological functions and behavior in higher animals. The generation and transmission of nerve impulses in vertebrates is mediated by an incredibly complicated neural network that connects every part of the organism with the brain—itself an interconnected array of as many as 10^{12} cells.

Despite their complexity and diversity, the nervous systems of higher organisms all possess common features and common mechanisms. Physical or chemical stimuli are recognized by specialized **receptor proteins** in the membranes of **excitable cells.** Conformational changes in the receptor protein result in a change in enzyme activity or a change in the permeability of the membrane. These changes are then propagated throughout the cell or from cell to cell in specific and reversible ways to carry information through the organism. This chapter describes the characteristics of excitable cells and the mechanisms by which these cells carry information at high speeds through an organism.

38.1 The Cells of Nervous Systems

Neurons and **neuroglia** (or **glial cells**) are cell types unique to nervous systems. The reception and transmission of nerve impulses are carried out by neurons, whereas glial cells serve protective and supportive functions. ("Neuroglia" could be translated as "nerve glue.") Glial cells differ from neurons in several ways. Glial cells do not possess axons or synapses and they retain the ability to divide throughout their life spans. Glial cells, which outnumber neurons by at least 10 to 1 in in most animals, may be of several types. Certain glial cells, such as **Schwann cells,** envelop and surround neurons, forming a protective sheath. Other glial cells are phagocytic in nature and remove cellular debris from nervous tissue. Glial cells also form linings in the cavities of the brain and around the spinal cord.

Neurons are distinguished from other cell types by their long cytoplasmic extensions or projections, called **processes** (Figure 38.1). Most neurons consist of three distinct regions: the **cell body** (called the **soma**), the **axon,** and the **dendrites.** The cell body is similar to most other somatic cells and contains the nucleus as well as other organelles, such as the endoplasmic reticulum and mitochondria. The axon is a long, thin structure extending from the cell body. Its primary function is to carry nerve impulses from the cell body to the cellular extremities or termini. Most axons are microscopic in diameter, although some are 1 meter or more in length. Axons of certain cells, such as those outside the central nervous system, are covered by two layers or sheaths, an outer **cellular sheath** and an inner **myelin sheath.** Both sheaths are derived from Schwann cells. The myelin sheath is produced by the winding of the Schwann cell membrane around the axon several times. The cellular sheath is produced by a layer of Schwann cells that wrap themselves once around the axon and myelin sheath. Gaps between Schwann cells along the axon's length are called the **nodes of Ranvier;** axons are not insulated in these regions. The axon ends in small structures called **synaptic terminals, synaptic knobs,** or **synaptic bulbs.** Dendrites are short, highly branched structures emanating from the cell body that receive neural impulses and transmit them to the cell body. The space between a synaptic knob on one neuron and a dendrite ending of an adjacent neuron is the **synapse** or **synaptic cleft.**

Three kinds of neurons are found in higher organisms (Figure 38.2): sensory neurons, interneurons, and motor neurons. **Sensory neurons** acquire sensory signals, either directly or from specific receptor cells, and pass this information along to either **interneurons** or **motor neurons.** Interneurons

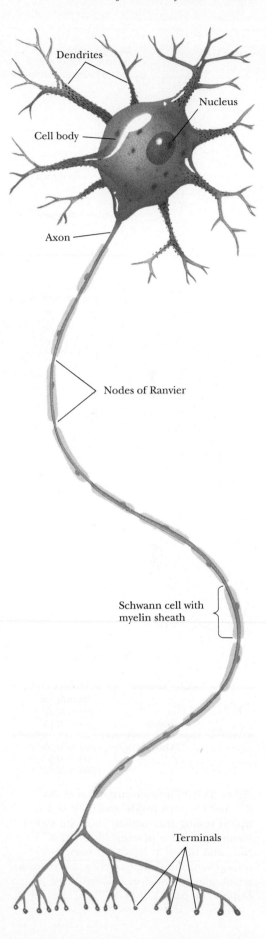

Figure **38.1** The anatomy of a mammalian motor neuron. The nucleus and most other organelles are contained in the cell body. One long axon and many shorter dendrites project from the body. The dendrites receive signals from other neurons and conduct them to the cell body. The axon transmits signals from this cell to other cells via the synaptic knobs. Glial cells called Schwann cells envelop the axon in layers of an insulating myelin membrane. Although glial cells lie in proximity to neurons in most cases, no specific connections (such as gap junctions, for example) connect glial cells and neurons. However, gap junctions can exist between adjacent glial cells.

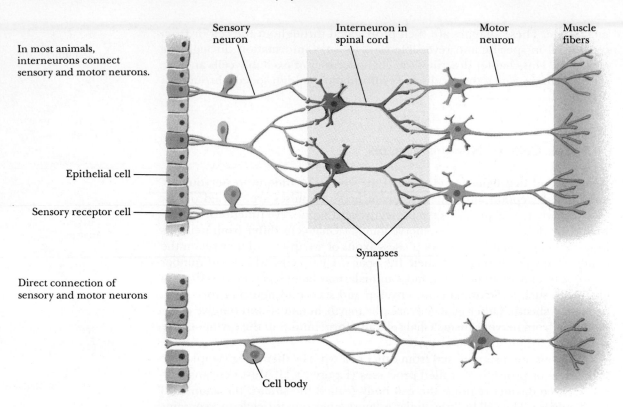

In most animals, interneurons connect sensory and motor neurons.

Epithelial cell

Sensory receptor cell

Direct connection of sensory and motor neurons

***Figure* 38.2** The connections between sensory neurons, interneurons, and motor neurons. The junctions between adjacent neurons are termed synapses.

simply pass signals from one neuron to another, whereas motor neurons pass signals from other neurons to muscle cells, thereby inducing muscle movement (motor activity). In some simpler organisms, interneurons are often absent, and sensory neurons are connected directly to motor neurons. In simple organisms such as sea anemones, even motor neurons are absent, and sensory neurons are connected directly to muscle cells.

38.2 Ion Gradients: Source of Electrical Potentials in Neurons

The impulses that are carried along axons, as signals pass from neuron to neuron, are electrical in nature. *These electrical signals occur as transient changes in the electrical potential differences (voltages) across the membranes of neurons (and other cells). Such potentials are generated by **ion gradients.*** The cytoplasm of a neuron at rest is low in Na^+ and Cl^- and high in K^+, relative to the extracellular fluid (Figure 38.3). These gradients are generated by the Na^+,K^+-ATPase (see Chapter 35). A resting neuron exhibits a potential difference of approximately -60 mV (that is, negative inside). Consider the potassium gradient in isolation (Figure 38.4). The electrochemical potential for an ion distributed across a membrane was given in Equation (35.4) as

$$\Delta G = G_2 - G_1 = RT \ln \left(\frac{C_2}{C_1} \right) + Z\mathcal{F}\Delta\psi$$

For the particular potassium gradient in Figure 38.3, what is the *equilibrium potential* at which no net ion flow would occur? At equilibrium, $\Delta G = 0$, and Equation (35.4) becomes

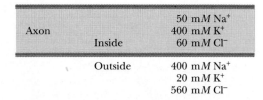

Axon		
		50 mM Na$^+$
	Inside	400 mM K$^+$
		60 mM Cl$^-$
	Outside	400 mM Na$^+$
		20 mM K$^+$
		560 mM Cl$^-$

***Figure* 38.3** The concentrations of Na$^+$, K$^+$, and Cl$^-$ ions inside and outside a typical resting mammalian axon are shown. Assuming relative permeabilities for K$^+$, Na$^+$, and Cl$^-$ of 1, 0.04, and 0.45, respectively, the Goldman equation yields a membrane potential of -60 mV. (See *A Deeper Look*, page 1221.)

A Deeper Look

The Actual Transmembrane Potential Difference

To appreciate the relationship between ion gradients and the actual potential difference across a nerve axon membrane, one must ask two questions: How can the *actual potential difference* in Figure 38.3 be calculated? and What do the equilibrium potentials for Na$^+$ and K$^+$ tell us with respect to the actual potential difference in Figure 38.3 of −60 mV? The answer to the first question was suggested by D. E. Goldman in 1943. Calculating a total potential for a situation like Figure 38.3, with several charged species distributed across a membrane, is more complicated than the case of a single ion, for which the Nernst equation is sufficient. Goldman

realized that, if the total potential is a consequence of the potentials for several ions, then none of the individual ions is likely to be *at equilibrium*. For systems not at equilibrium, *the **permeability** of the membrane to the various ions is just as important as the concentrations themselves.* Goldman's equation, which describes a **steady state** rather than an equilibrium, is

$$\Delta\psi = \frac{RT}{\mathscr{F}} \ln\left(\frac{\Sigma P_C[C]_{outside} + \Sigma P_A[A]_{inside}}{\Sigma P_C[C]_{inside} + \Sigma P_A[A]_{outside}}\right)$$

where [C] and [A] refer to cation and anion concentrations, respectively, and P_C and P_A are the respective permeability coefficients (Chapter 35) of cations and anions. Applying Goldman's equation to the concentrations and permeabilities shown in Figure 38.3 yields a

value for $\Delta\psi$ of −60 mV, in agreement with values measured experimentally in neurons.

The answer to the second question is fundamental to the generation of nerve impulses. The equilibrium potential for Na$^+$ (+53.4 mV) is very far from the typical measured potential, whereas the equilibrium potential for K$^+$ is very close to the observed value. This means that Na$^+$ ions have a much larger thermodynamic tendency to move into the neurons than K$^+$ ions have to move outward. If the axon membrane were suddenly made permeable to these ions, the predominant event would be a massive influx of Na$^+$, together with a less dramatic efflux of K$^+$—just what happens during the conduction of a nerve impulse.

$$RT \ln\left(\frac{C_2}{C_1}\right) = -Z\mathscr{F}\Delta\psi \qquad (38.1)$$

This is a form of the **Nernst equation,** which relates concentration differences and potential differences. For the case in Figure 38.4, with $Z = 1$,

$$RT \ln\left(\frac{400}{20}\right) = -\mathscr{F}\Delta\psi \qquad (38.2)$$

and, at 25° C, with $\mathscr{F} = 96.48$ kJ/V-mol,

$$\Delta\psi = -77 \text{ mV}$$

At a potential difference of −77 mV, the net flow of K$^+$ across this membrane is zero. What is the equilibrium potential for Na$^+$ in Figure 38.3? Applying the Nernst equation again, an equilibrium potential of +53.4 mV for Na$^+$ in Figure 38.3 can be calculated.

38.3 The Action Potential

Nerve impulses, also called **action potentials,** are transient changes in the membrane potential that move rapidly along nerve cells. Action potentials are created when the membrane is locally **depolarized** by approximately 20 mV— from the resting value of about −60 mV to a new value of approximately −40 mV. This small change is enough to have a dramatic effect on specific proteins in the axon membrane called **voltage-gated ion channels.** These proteins are ion channels that are specific either for Na$^+$ or K$^+$. These ion channels are normally closed at the resting potential of −60 mV. When the poten-

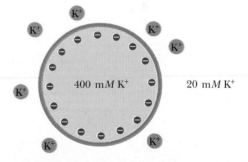

Figure 38.4 The K$^+$ concentration inside a resting axon is approximately 400 mM, whereas the concentration outside is 20 mM. In the absence of an electrical potential difference across the membrane, there would be a strong thermodynamic tendency for K$^+$ to flow out of the axon. According to the Nernst equation, the concentration gradient of K$^+$ would be exactly balanced (at 25°C) by a potential difference of −77 mV (negative inside). If $\Delta\psi = -77$ mV, the net flow of K$^+$ would be zero.

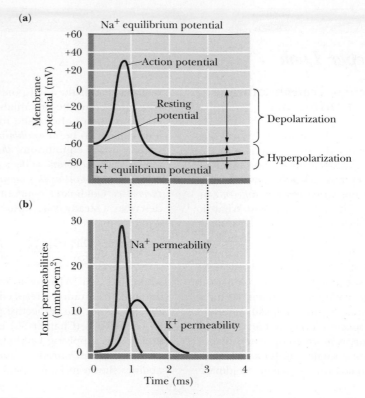

***Figure* 38.5**   The time dependence of an axon potential, compared with the ionic permeabilities of Na^+ and K^+. (a) The rapid rise in membrane potential from -60 mV to slightly more than $+30$ mV is referred to as a "depolarization." This depolarization is caused (b) by a sudden increase in the permeability of Na^+. As the Na^+ permeability decreases, K^+ permeability is increased and the membrane potential drops, eventually falling below the resting potential—a state of "hyperpolarization"—followed by a slow return to the resting potential. *(Adapted from Hodgkin, A., and Huxley, A., 1952. A quantitative description of membrane current and its application to conduction and excitation in nerve. Journal of Physiology 117:500–544.)*

tial difference rises to -40 mV, the "gates" of the Na^+ channels are opened and Na^+ ions begin to flow into the cell, as expected from the electrochemical potential exerted on Na^+. As Na^+ enters the cell, the membrane potential continues to increase, and additional Na^+ channels are opened (Figure 38.5). The potential rises to more than $+30$ mV. At this point, Na^+ influx slows and stops, because the actual potential is approaching the Na^+ equilibrium potential. As the Na^+ channels close, K^+ channels begin to open and K^+ ions stream out of the cell, returning the membrane potential to negative values. The potential eventually overshoots its resting value a bit. At this point, K^+ channels close and the resting potential is eventually restored by action of the Na^+,K^+-ATPase and the other channels. These transient increases and decreases, first in Na^+ permeability and then in K^+ permeability, were first observed by Alan Hodgkin and Andrew Huxley. For this and related work, Hodgkin and Huxley, along with J. C. Eccles, won the Nobel Prize in physiology or medicine in 1963.

The Action Potential Is Mediated by the Flow of Na^+ and K^+ Ions

These changes in potential in one part of the axon are rapidly passed along the axonal membrane (Figure 38.6). The sodium ions that rush into the cell in one localized region actually diffuse farther along the axon, raising the Na^+ concentration *and* depolarizing the membrane, causing Na^+ gates to

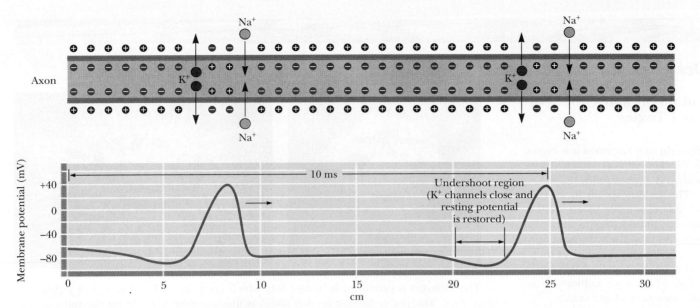

Figure 38.6 The propagation of action potentials along an axon. Figure 38.5 showed the time dependence of an action potential at a discrete point on the axon. This figure shows how the membrane potential varies along the axon as an action potential is propagated. (For this reason, the shape of the action potential is the apparent reverse of that shown in Figure 38.5.) At the leading edge of the action potential, membrane depolarization causes Na^+ channels to open briefly. As the potential moves along the axon, the Na^+ channels close and K^+ channels open, leading to a drop in potential and the onset of hyperpolarization. When the resting potential is restored, another action potential can be initiated.

open in that adjacent region of the axon. In this way, the action potential moves down the axon in wavelike fashion. This simple process has several very dramatic properties:

1. Action potentials propagate very rapidly—up to and sometimes exceeding 100 meters per second.

2. The action potential is not attenuated (diminished in intensity) as a function of distance transmitted.

The input of energy all the way along an axon—in the form of ion gradients maintained by Na^+,K^+-ATPase—ensures that the shape and intensity of the action potential will be maintained over long distances. The action potential has an all-or-none character. There are no gradations of amplitude; a given neuron is either at rest (with a polarized membrane), or is conducting a nerve impulse (with a reversed polarization). Because nerve impulses display no variation in amplitude, the size of the action potential is not important in processing signals in the nervous system. *Instead, it is the number of action potential firings and the frequency of firing that carry specific information.*

38.4 The Voltage-Gated Sodium and Potassium Channels

The action potential is a delicately orchestrated interplay between the Na^+,K^+-ATPase and the voltage-gated Na^+ and K^+ channels, initiated by a stimulus at the postsynaptic membrane. The density and distribution of Na^+ channels along the axon are different for myelinated and unmyelinated axons (Figure 38.7). In unmyelinated axons, Na^+ channels are uniformly distrib-

A Deeper Look

Tetrodotoxin and Other Na⁺ Channel Toxins

Tetrodotoxin and **saxitoxin** are highly specific blockers of Na⁺ channels and bind with very high affinity ($K_D \cong$ 1 nM). This unique specificity and affinity has made it possible to use radioactive forms of these toxins to purify Na⁺ channels and map their distribution on axons. Tetrodotoxin is found in the skin and several internal organs of puffer fish, also known as blowfish or swellfish, members of the family *Tetraodontidae,* which react to danger by inflating themselves with water or air to nearly spherical (and often comical) shapes (see figure). Although tetrodotoxin poisoning can easily be fatal, puffer fish are delicacies in Japan, where they are served in a dish called *fugu.* For this purpose the puffer fish must be cleaned and prepared by specially trained chefs. Saxitoxin is made by *Gonyaulax catenella* and *G. tamarensis,* two species of marine dinoflagellates or plankton that are responsible for "red

Tetrodotoxin is produced by puffer fish, which are prepared and served in Japan as *fugu.* The puffer fish on the left above is unexpanded; the one on the right is inflated.

tides" that cause massive fish kills. Saxitoxin is concentrated by certain species of mussels, scallops, and other shellfish that are exposed to red tides. Consumption of these shellfish by animals, including humans, can be fatal. In addition to these toxins, which prevent the Na⁺ channel from opening, there are equally poisonous agents that block the Na⁺ channel in an open state, permitting Na⁺ to stream into the cell without control, destroying the Na⁺ gradients. Included in this group of toxins (Figure 38.8) are **veratridine** from *Schoenocaulon officinalis,* a member of the lily family, and **batrachotoxin,** a compound found in skin secretions of a Colombian frog, *Phyllobates aurotaenia.* These skin secretions have traditionally been used as arrow poisons.

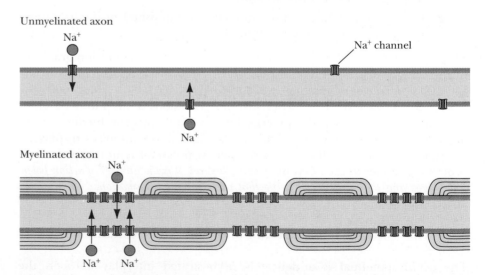

Figure **38.7**   Na⁺ channels are infrequently and randomly distributed in unmyelinated nerve. In myelinated axons, Na⁺ channels are clustered in large numbers in the nodes of Ranvier, between the regions surrounded by myelin sheath structures.

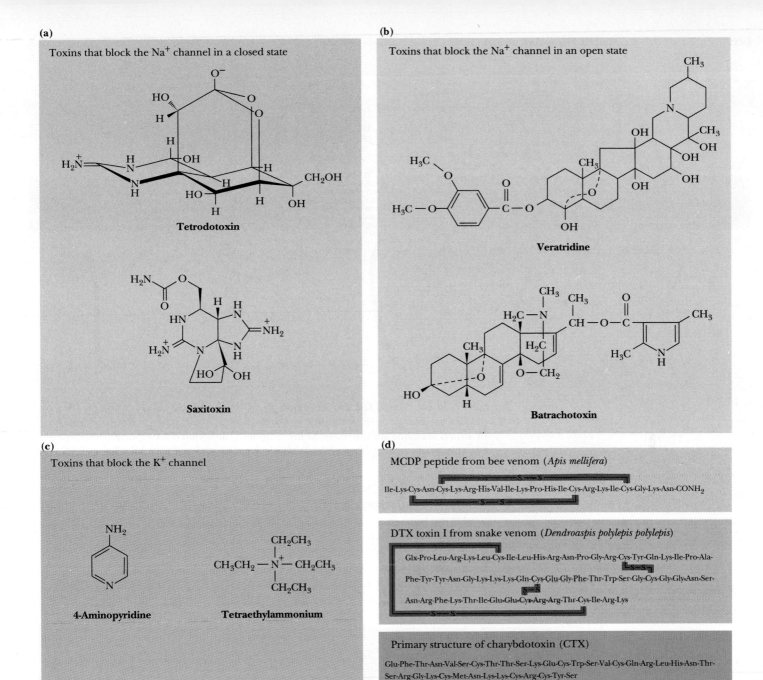

(a)

Toxins that block the Na⁺ channel in a closed state

Tetrodotoxin

Saxitoxin

(b)

Toxins that block the Na⁺ channel in an open state

Veratridine

Batrachotoxin

(c)

Toxins that block the K⁺ channel

4-Aminopyridine

Tetraethylammonium

(d)

MCDP peptide from bee venom (*Apis mellifera*)

Ile-Lys-Cys-Asn-Cys-Lys-Arg-His-Val-Ile-Lys-Pro-His-Ile-Cys-Arg-Lys-Ile-Cys-Gly-Lys-Asn-CONH₂

DTX toxin I from snake venom (*Dendroaspis polylepis polylepis*)

Glx-Pro-Leu-Arg-Lys-Leu-Cys-Ile-Leu-His-Arg-Asn-Pro-Gly-Arg-Cys-Tyr-Gln-Lys-Ile-Pro-Ala-
Phe-Tyr-Tyr-Asn-Gly-Lys-Lys-Lys-Gln-Cys-Glu-Gly-Phe-Thr-Trp-Ser-Gly-Cys-Gly-Gly-Asn-Ser-
Asn-Arg-Phe-Lys-Thr-Ile-Glu-Glu-Cys-Arg-Arg-Thr-Cys-Ile-Arg-Lys

Primary structure of charybdotoxin (CTX)

Glu-Phe-Thr-Asn-Val-Ser-Cys-Thr-Thr-Ser-Lys-Glu-Cys-Trp-Ser-Val-Cys-Gln-Arg-Leu-His-Asn-Thr-
Ser-Arg-Gly-Lys-Cys-Met-Asn-Lys-Lys-Cys-Arg-Cys-Tyr-Ser

Figure 38.8 Effectors of Na⁺ channels include (a) tetrodotoxin and saxitoxin, which block the Na⁺ channel in a closed state, and (b) veratridine and batrachotoxin, which block the Na⁺ channel in an open state. K⁺-channel blockers include (c) 4-aminopyridine, tetraethylammonium ion, mast cell degranulating peptide (MCDP), dendrotoxin (DTX), and charybdotoxin (CTX). See *A Deeper Look*, page 1227.

uted, although they are few in number—approximately 20 channels per μm^2. On the other hand, in myelinated axons, Na⁺ channels are **clustered** at the nodes of Ranvier. In these latter regions, they occur with a density of approximately 10,000 per μm^2. Elucidation of these distributions of Na⁺ channels was made possible by the use of several Na⁺-channel toxins (Figure 38.8).

The purified Na⁺ channel from mammalian brain is a heterotrimer consisting of α- (260 kD), β_1- (36 kD), and β_2- (33 kD) subunits (Figure 38.9). All three subunits are exposed to the extracellular surface and are heavily glycosylated. The β_2-subunit is attached to the α-subunit by disulfide bonds. The 260-kD α-subunit contains the binding site for toxins. The Na⁺-channel

Na⁺ channel

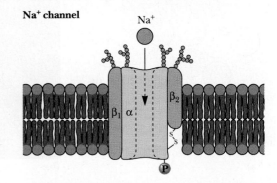

Figure 38.9 The Na⁺ channel comprises three subunits, denoted α, β_1, and β_2. A disulfide bridge links α and β_2 as shown. All three subunits are glycosylated and the α-subunit can be phosphorylated on the cytoplasmic surface.

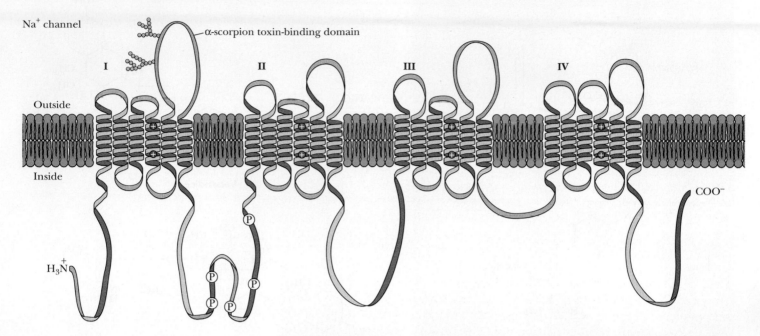

Figure 38.10   Model for the arrangement of the Na⁺-channel α-subunit in the plasma membrane. The α-subunit consists of four domains (I through IV), each of which contains six transmembrane α-helices, designated S1 through S6. Phosphorylation sites (P) and locations of positive charges on helix S4 are indicated.

α-subunit contains four domains of 300 to 400 amino acids each (Figure 38.10), with an approximately 50% identity or conservation in their amino acid sequences. Each domain contains six regions (denoted S1 to S6) of probable α-helical structure, which are long enough to be membrane-spanning segments. Segments S5 and S6 are uniformly hydrophobic, S1 and S2 are hydrophobic with an occasional hydrophilic residue, and S3 has several charged residues. S4 has both hydrophobic and positively charged residues; its sequence is highly conserved among voltage-gated Na⁺, K⁺, and Ca²⁺

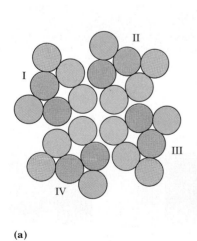

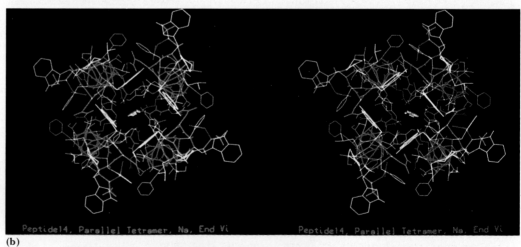

Figure 38.11   (a) A model for the formation of an ion channel from the four membrane-spanning domains of the Na⁺ channel. The ion channel is located in the center of the structure. Four corresponding helical segments (purple), one from each domain, form the wall of the channel. Voltage-dependent gating is thought to be mediated by the S4 segments (green). (b) A stereo molecular graphic of the four helical segments that form the pore.

(Photo kindly provided by Mauricio Montal, University of California, San Diego.)

A Deeper Look

Potassium Channel Toxins

K⁺-channel blockers include (Figure 38.8) **4-aminopyridine, tetraethylammonium ion,** and several peptide toxins, including the **dendrotoxins (DTX), mast cell degranulating peptide (MCDP),** and **charybdotoxin (CTX).** Dendrotoxin I is a 60-residue peptide from *Dendroaspis polylepsis,* the dangerous black mamba snake of sub-Saharan Africa. MCDP, a bee venom toxin that has a degranulating action on mast cells, is a potent convulsant. It is a 22-residue peptide with two disulfide bonds, one proline, and a C-terminal amide. Charybdotoxin is a minor component of the venom of the scorpion, *Leiurus quinquestriatus.* It is a 37-residue peptide with 8 positively charged residues (3 arginines, 4 lysines, and a histidine). All these agents bind with high affinity to membranes containing voltage-activated K⁺ channels.

---

channels. Nearly every third residue in S4 segments is a lysine or an arginine. Studies with mutant forms of the Na⁺ channel have provided evidence that this segment may be part of the voltage-sensing mechanism in the Na⁺ channel. Shosaku Numa and his collaborators have produced Na⁺ channels with one or more of the positively charged lysines and arginines of S4 missing, and found that a reduction in the net positive charge of S4 results in a reduction in the steepness of the voltage dependence of sodium channel activation. S4 senses the transmembrane electric field, thereby controlling the gating of the sodium channel. Numa has suggested that the four S2 segments of an α-subunit form the walls of the Na⁺ channel (Figure 38.11).

Voltage-gated potassium channels have now been identified in many tissues and species. Although the primary structures of these proteins are highly homologous, they display a variety of functional properties. Just as for the voltage-gated sodium channels, the high-affinity binding of several specific K⁺-channel blockers has aided in the identification and characterization of K⁺ channels. The K⁺ channel from rat synaptosomal membranes consists of an α-subunit of 76 to 80 kD and a β-subunit of 38 kD. Phosphorylation of the α-subunit, either by a cAMP-dependent kinase or by an endogenous protein kinase, leads to activation of the K⁺ channel. The α-subunit possesses binding sites for dendrotoxin I and for MCDP (Figure 38.8). The K⁺-channel α-subunit from *Drosophila* consists of 616 amino acids. Hydropathy analysis reveals six or seven sequences with hydrophobic character and helix-forming potential. The fourth of the six putative transmembrane segments has striking homology with S4 of the sodium channel, and the proposed arrangement of the K⁺-channel protein in Figure 38.12 is similar to that for each of the four homologous domains of the sodium channel. The role of S4 of the sodium channel as a voltage sensor and its similarity to the fourth segment of the K⁺ channel raise the question of whether this fourth segment might play a similar voltage-sensing role in the K⁺ channel.

38.5 Cell–Cell Communication at the Synapse

How are neuronal signals passed from one neuron to the next? Neurons are juxtaposed at the synapse (Figure 38.2). The space between the two neurons is called the **synaptic cleft.** The number of synapses in which any given neuron is involved (Figure 38.13) varies greatly. There may be as few as one synapse per postsynaptic cell (in the midbrain) to many thousands per cell. Typically, 10,000 synaptic knobs may impinge a single spinal motor neuron,

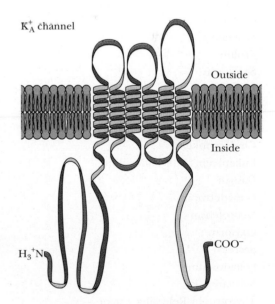

***Figure* 38.12** A model for the insertion of the potassium channel α-subunit in the plasma membrane. The transmembrane domain consists of six putative membrane-spanning helical segments and is highly homologous with the corresponding domains of the Na⁺ channel.

Figure **38.13** Synaptic junctions on a typical motor neuron. The motor neuron has dendrites, an axon, and a nucleus. Synaptic knobs from other neurons impinge on the dendrites and the cell body of the motor neuron. The axon is insulated by myelin sheath structures.

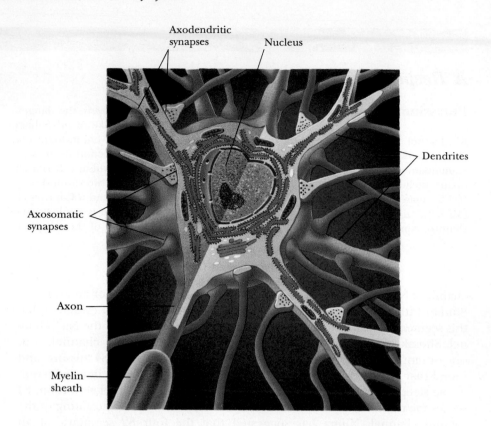

Table **38.1**
Families of Neurotransmitters

Cholinergic Agents
Acetylcholine

Catecholamines
Norepinephrine (noradrenaline)
Epinephrine (adrenaline)
L-Dopa
Dopamine
Octopamine

Amino Acids (and Derivatives)
γ-Aminobutyric acid (GABA)
Alanine
Aspartate
Cystathione
Glycine
Glutamate
Histamine
Proline
Serotonin
Taurine
Tyrosine

Peptide Neurotransmitters
Cholecystokinin
Enkephalins and endorphins
Gastrin
Gonadotropin
Neurotensin
Oxytocin
Secretin
Somatostatin
Substance P
Thyrotropin Releasing Factor
Vasopressin
Vasoactive Intestinal Peptide (VIP)

Gaseous Neurotransmitters
Carbon monoxide (CO)
Nitric oxide (NO)

with 8000 on the dendrites and 2000 on the soma or cell body. The ratio of synapses to neurons in the human forebrain is approximately 40,000 to 1!

Synapses are actually quite specialized structures and there are several different types. *A few synapses in mammals, termed* **electrical synapses,** *are characterized by a very small gap—approximately 2 nm—between the* **presynaptic cell** *(which delivers the signal) and the* **postsynaptic cell** *(which receives the signal).* At electrical synapses, the arrival of an action potential on the presynaptic membrane leads directly to depolarization of the postsynaptic membrane, initiating a new action potential in the postsynaptic cell. However, most synaptic clefts are much wider—on the order of 20 nm to 50 nm. *Here, an action potential in the presynaptic membrane causes secretion of a chemical substance—called a* **neurotransmitter**—*by the presynaptic cell.* This substance binds to receptors on the postsynaptic cell, initiating a new action potential. Synapses of this type are thus **chemical synapses.**

Different synapses utilize specific neurotransmitters. The **cholinergic synapse,** a paradigm for chemical transmission mechanisms at synapses, employs acetylcholine as a neurotransmitter. Other important neurotransmitters and receptors have been discovered and characterized. These all fall into one of several major classes, including **amino acids** (and their derivatives), **catecholamines, peptides,** and **gaseous neurotransmitters.** Table 38.1 lists some, but not all, of the known neurotransmitters.

The Cholinergic Synapses

In **cholinergic synapses,** small **synaptic vesicles** inside the synaptic knobs contain large amounts of acetylcholine (approximately 10,000 molecules per vesicle; Figure 38.14). When the membrane of the synaptic knob is stimulated by an arriving action potential, special **voltage-gated Ca²⁺ channels** open and Ca²⁺ ions stream into the synaptic knob, causing the acetylcholine-containing

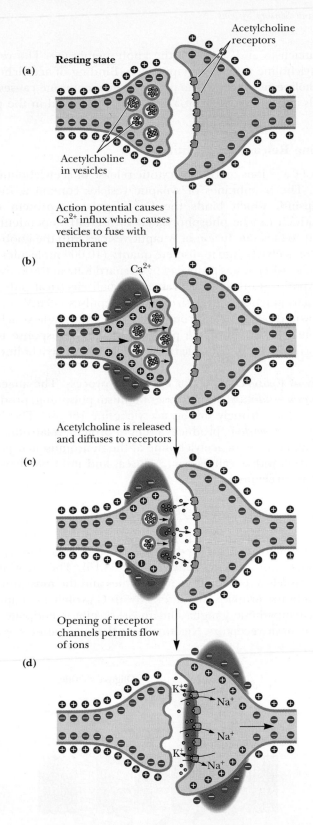

Resting state
(a)

Acetylcholine receptors

Acetylcholine in vesicles

Action potential causes Ca^{2+} influx which causes vesicles to fuse with membrane

(b)

Ca^{2+}

Acetylcholine is released and diffuses to receptors

(c)

Opening of receptor channels permits flow of ions

(d)

K$^+$

Na$^+$

Na$^+$

Na$^+$

K$^+$

Na$^+$

***Figure* 38.14** Cell–cell communication at the synapse (a) is mediated by neurotransmitters such as acetylcholine, produced from choline by choline acetyltransferase. The arrival of an action potential at the synaptic knob (b) opens Ca^{2+} channels in the presynaptic membrane. Influx of Ca^{2+} induces the fusion of acetylcholine-containing vesicles with the plasma membrane and release of acetylcholine into the synaptic cleft (c). Binding of acetylcholine to receptors in the postsynaptic membrane opens Na$^+$ channels (d). The influx of Na$^+$ depolarizes the postsynaptic membrane, generating a new action potential.

Nicotiana tabacum

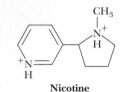

Nicotine

Amanita muscaria

Muscarine

Figure 38.15 Two types of acetylcholine receptors are known. Nicotinic acetylcholine receptors are locked in the open conformation by nicotine. Obtained from tobacco plants, nicotine is named for Jean Nicot, French ambassador to Portugal, who sent tobacco seeds to France in 1550 for cultivation. Muscarinic acetylcholine receptors are stimulated by muscarine, obtained from the intensely poisonous mushroom *Amanita muscaria*.

vesicles to attach to and fuse with the knob membrane. The vesicles open, spilling acetylcholine into the synaptic cleft. Binding of acetylcholine to specific **acetylcholine receptors** in the postsynaptic membrane causes opening of ion channels and the creation of a new action potential in the postsynaptic neuron.

Acetylcholine Release Is Quantized

The action of Ca^{2+} ions on this exocytotic release of acetylcholine is not fully understood. The membranes of synaptic vesicles contain a 75-kD protein called **synapsin-I**, which binds the Ca^{2+}-regulatory protein calmodulin. Synapsin-I, which can be phosphorylated by a cAMP-dependent protein kinase, appears to facilitate fusion of synaptic vesicles with the knob membrane. The release of acetylcholine in discrete quanta (10,000 molecules per vesicle) was a novel idea when it was proposed by Bernard Katz in the early 1950s. Katz based his hypothesis on the occurrence of small electrical pulses in resting neuromuscular junctions. The size of these pulses—3 mV and less—was about what would be expected from the release of a single vesicle's contents of acetylcholine. The size of the pulses observed in response to an action potential corresponds to the release of up to 400 acetylcholine-containing vesicles.

A variety of toxins can alter or affect this process. The anaerobic bacterium *Clostridium botulinum*, which causes botulism poisoning, produces several toxic proteins that strongly inhibit acetylcholine release. The black widow spider, *Lactrodectus mactans*, produces a venom protein, **α-latrotoxin**, that stimulates abnormal release of acetylcholine at the neuromuscular junction. The bite of the black widow causes pain, nausea, and mild paralysis of the diaphragm but is rarely fatal.

Two Classes of Acetylcholine Receptors

Two different acetylcholine receptors are found in postsynaptic membranes. They were originally distinguished by their responses to **muscarine**, a toxic alkaloid in toadstools, and **nicotine** (Figure 38.15). The **nicotinic receptors** are cation channels in postsynaptic membranes and the **muscarinic receptors** are transmembrane proteins that interact with G proteins (Chapter 37). The receptors in sympathetic ganglia and those in motor endplates of skeletal muscle are nicotinic receptors. Nicotine locks the ion channels of these recep-

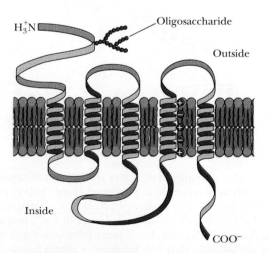

Figure 38.16 A model for the arrangement of the acetylcholine receptor α-subunit in the postsynaptic membrane.

(a)

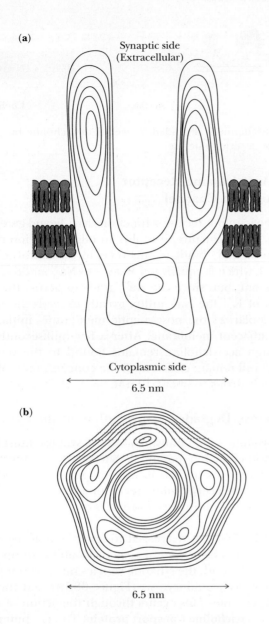

Synaptic side
(Extracellular)

Cytoplasmic side

← 6.5 nm →

(b)

← 6.5 nm →

***Figure* 38.17** (a) Side view and (b) top view of the structure of the nicotinic acetylcholine receptor Na$^+$ channel, as determined by image reconstruction from electron micrographs. The diameter of the pore at the mouth of the channel (facing the synapse) is 2.2 nm.

(Adapted from Brisson, A., and Unwin, P. N. T., 1985. Quaternary structure of the acetylcholine receptor. Nature 315:474–477.)

tors in their open conformation. Muscarinic receptors are found in smooth muscle and in glands. Muscarine mimics the effect of acetylcholine on these latter receptors.

The nicotinic acetylcholine receptor is a transmembrane glycoprotein with an approximate molecular mass of 270 kD, consisting of four different subunits, α (54 kD), β (56 kD), γ (58 kD), and δ (60 kD), with a quaternary structure of $\alpha_2\beta\gamma\delta$. Each α-subunit possesses a binding site for acetylcholine. The four different subunits have homologous sequences and may have evolved via gene duplication. Each subunit includes five hydrophobic regions that are postulated to be helical, membrane-spanning segments (Figure 38.16). Three-dimensional reconstructions indicate that the receptor has a cylindrical shape, with approximate fivefold symmetry and a central pore (Figure 38.17). Models indicate that several charged residues are clustered on one face of the fourth helix on each subunit. These charged helical faces may line the wall of the transmembrane channel.

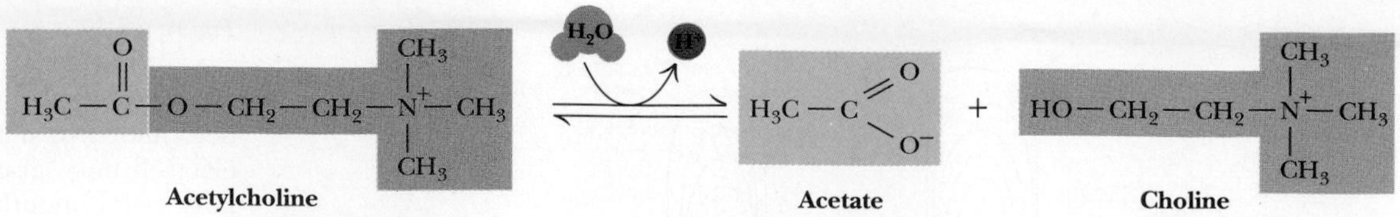

Figure 38.18 Acetylcholine is degraded to acetate and choline by acetylcholinesterase, a serine esterase.

The Nicotinic Acetylcholine Receptor Is a Ligand-Gated Ion Channel

The nicotinic acetylcholine receptor functions as a **ligand-gated ion channel,** and, on the basis of its structure, it is also an **oligomeric ion channel.** When acetylcholine (the ligand) binds to this receptor, a conformational change opens the channel, which is equally permeable to Na^+ and K^+. Na^+ rushes in while K^+ streams out, but, since the Na^+ gradient across this membrane is steeper than that of K^+, the Na^+ influx greatly exceeds the K^+ efflux. The influx of Na^+ depolarizes the postsynaptic membrane, initiating an action potential in the adjacent membrane. After a few milliseconds, the channel closes, even though acetylcholine remains bound to the receptor. At this point the channel will remain closed until the concentration of acetylcholine in the synaptic cleft drops to about 10 nM.

Acetylcholinesterase Degrades Acetylcholine in the Synaptic Cleft

Following every synaptic signal transmission, the synapse must be readied for the arrival of another action potential. Several things must happen very quickly. First, the acetylcholine left in the synaptic cleft must be rapidly degraded to resensitize the acetylcholine receptor and to restore the excitability of the postsynaptic membrane. This reaction is catalyzed by **acetylcholinesterase** (Figure 38.18).

When [acetylcholine] has decreased to low levels, acetylcholine dissociates from the receptor, which thereby regains its ability to open in a ligand-dependent manner. Second, the synaptic vesicles must be reformed from the presynaptic membrane by endocytosis (Figure 38.19) and then must be restocked with acetylcholine. This occurs through the action of an ATP-driven H^+ pump and an **acetylcholine transport protein.** The H^+ pump in this case is a member of the family of **V-type ATPases.** It uses the free energy of ATP hydrolysis to create an H^+ gradient across the vesicle membrane. This gradient is used by the acetylcholine transport protein to drive acetylcholine into the vesicle, as shown in Figure 38.19.

Antagonists of the nicotinic acetylcholine receptor are particularly potent neurotoxins. These agents, which bind to the receptor and prevent opening of the ion channel, include **d-tubocurarine** (Figure 38.20), the active agent in the South American arrow poison **curare,** and several small proteins from poisonous snakes. These latter agents include **cobratoxin** from cobra venom, and **α-bungarotoxin,** from *Bungarus multicinctus,* a snake common in Taiwan.

Muscarinic Receptor Function Is Mediated by G Proteins

There are several different types of muscarinic acetylcholine receptors, with different structures and different apparent functions in synaptic transmission. However, certain structural and functional features are shared by this class of receptors. Muscarinic receptors that have been isolated or cloned and sequenced are 70-kD glycoproteins (of which about 25% to 27% are carbohydrates) and are members of the 7-transmembrane segment (7-TMS) family of receptors (Chapter 37).

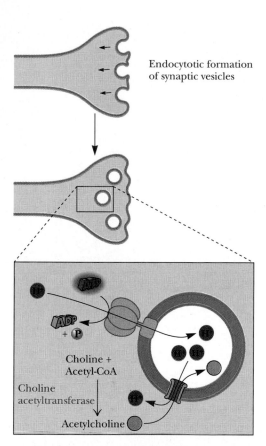

Figure 38.19 Following a synaptic transmission event, acetylcholine is repackaged in vesicles in a multistep process. Synaptic vesicles are formed by endocytosis, and acetylcholine is synthesized by choline acetyltransferase. A proton gradient is established across the vesicle membrane by an H^+-translocating ATPase, and a proton-acetylcholine transport protein transports acetylcholine into the synaptic vesicles, exchanging acetylcholine for protons in an electrically neutral antiport process.

Chondrodendron

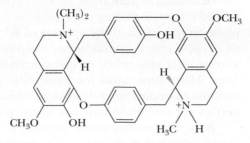

Tubocurarine

Deadly nightshade
(*Atropa belladonna*)

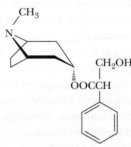

Atropine

Indian cobra
(*Naja naja*)

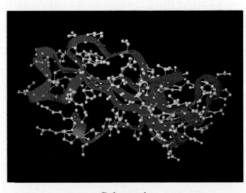

Cobratoxin

Bungarus multicinctus

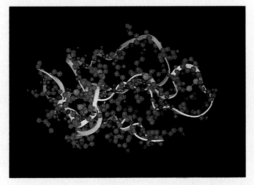

α-Bungarotoxin

***Figure* 38.20** Tubocurarine, obtained from the plant *Chondrodendron tomentosum,* is the active agent in "tube curare," named for the bamboo tubes in which it is kept by South American tribal hunters. Atropine is produced by *Atropa belladonna,* the deadly nightshade. Cobratoxin and α-bungarotoxin are produced by the cobra (*Naja naja*) and the banded krait snake (*Bungarus multicinctus*), respectively.

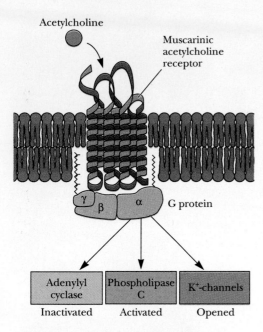

Acetylcholine

Muscarinic acetylcholine receptor

γ β α G protein

Adenylyl cyclase	Phospholipase C	K⁺-channels
Inactivated	Activated	Opened

Figure 38.21 Muscarinic acetylcholine receptors are typical 7-transmembrane segment receptor proteins. Binding of acetylcholine to these receptors activates G proteins, which inactivate adenylyl cyclase, activate phospholipase C, and open K⁺ channels.

Activation of muscarinic receptors (by binding of acetylcholine) results in several effects, including the inhibition of **adenylyl cyclase,** the stimulation of **phospholipase C,** and the **opening of K⁺ channels.** As shown in Figure 38.21, all of these effects of muscarinic receptors are mediated by G proteins, and it appears that muscarinic receptor activation involves several different G proteins. Many antagonists for muscarinic acetylcholine receptors are known, including **atropine** from *Atropa belladonna,* the deadly nightshade plant, whose berries are sweet and tasty but highly poisonous (Figure 38.20).

Both the nicotinic and muscarinic acetylcholine receptors are sensitive to certain agents that, in effect, overstimulate the receptor. This effect can happen in two ways. Certain substances, such as the cations **decamethonium** and **succinylcholine** (Figure 38.22) bind to (and activate) the receptors but are not rapidly degraded by acetylcholinesterase. Thus, they remain bound to the receptors for long periods, keeping the ion channel in the open conformation and preventing the reestablishment of receptor sensitivity. Similar effects can be produced by agents that inactivate acetylcholinesterase itself. Acetylcholinesterase is a serine esterase similar to trypsin and chymotrypsin (Chapter 13). The reactive serine at the active site of such enzymes is a vulnerable target for organophosphorus inhibitors (Figure 38.23). **DIPF** and related agents form stable covalent complexes with the active-site serine, irreversibly blocking the enzyme. **Malathion** and **parathion** are commonly used insecticides, and **sarin** and **tabun** are nerve gases used in chemical warfare. All these agents effectively block nerve impulses, stop breathing, and cause death by suffocation.

Milder inhibitors of the acetylcholinesterase reaction are useful therapeutic agents. **Physostigmine,** an alkaloid found in calabar beans, and **neostigmine,** a synthetic analog, contain carbamoyl ester groups (Figure 38.23). Reaction with the active-site serine of acetylcholinesterase leaves an intermediate that is hydrolyzed only very slowly, effectively inhibiting the enzyme. Physostigmine and neostigmine have been used to treat **myasthenia gravis,** a chronic disorder that causes muscle weakness after the muscle has been exercised, and eventual paralysis. Myasthenia gravis is an autoimmune disease, in which individuals produce antibodies that bind to their own acetylcholine receptors, blocking the response to acetylcholine. By blocking acetylcholinesterase (thus allowing acetylcholine levels in the synaptic cleft to remain high), physostigmine and neostigmine can suppress the symptoms of myasthenia gravis.

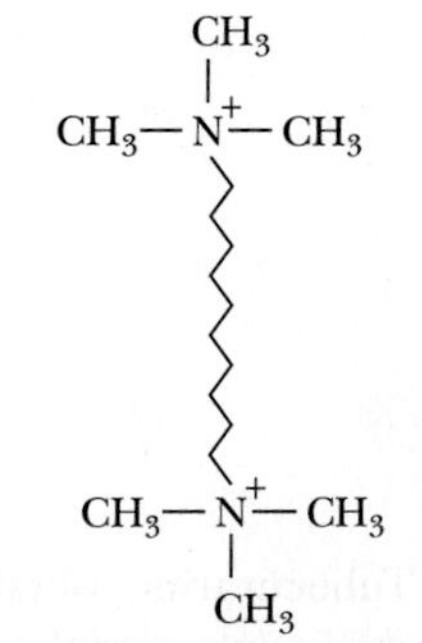

38.6 Other Neurotransmitters and Synaptic Junctions

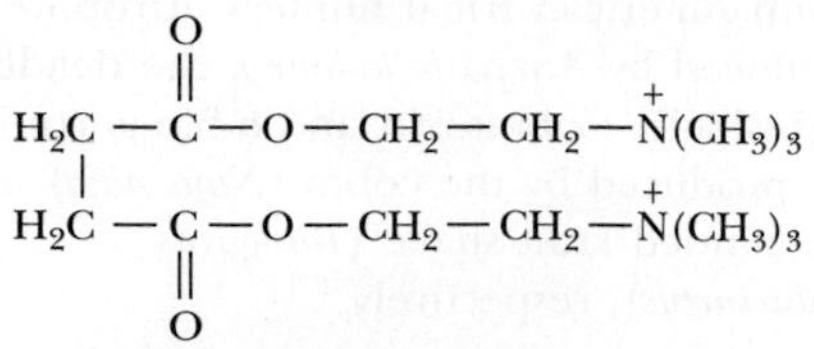

Synaptic junctions that use amino acids, catecholamines, and peptides (see Table 38.1) appear to operate much the way the cholinergic synapses do. Presynaptic vesicles release their contents into the synaptic cleft, where the neurotransmitter substance can bind to specific receptors on the postsynaptic membrane to induce a conformation change and elicit a particular response. This common mechanism notwithstanding, the synapses using these various neurotransmitters display markedly different properties. Some of these respond quickly to neuronal signals, whereas others respond slowly. Some of these neurotransmitters are **excitatory** in nature and stimulate postsynaptic neurons to transmit impulses, whereas others are **inhibitory** and prevent the postsynaptic neuron from carrying other signals. Moreover, it is also becoming clear that *each known neurotransmitter acts on a **family** of postsynaptic receptors.* Just as acetylcholine acts both on nicotinic and muscarinic receptors, so most of the known neurotransmitters act on several (and in some cases, many) different kinds of receptors. Biochemists are just beginning to understand the sophistication and complexity of neuronal signal transmission.

CH₃

CH₃—N⁺—CH₃

CH₃—N⁺—CH₃

CH₃

Decamethonium

$$H_2C - \overset{O}{\overset{\|}{C}} - O - CH_2 - CH_2 - \overset{+}{N}(CH_3)_3$$
$$H_2C - \underset{O}{\underset{\|}{C}} - O - CH_2 - CH_2 - \overset{+}{N}(CH_3)_3$$

Succinylcholine

Figure 38.22 The structures of decamethonium ion and succinylcholine.

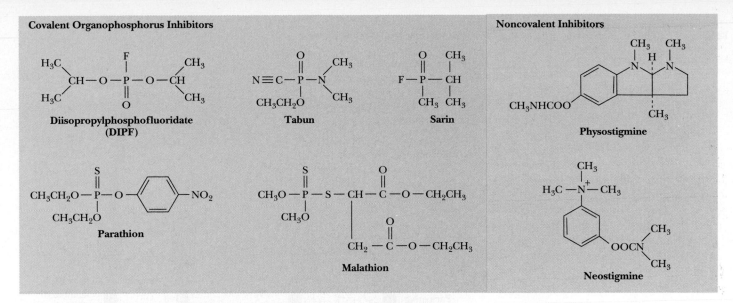

Figure 38.23 Covalent inhibitors (green) of acetylcholinesterase include DIPF, the nerve gases tabun and sarin, and the insecticides parathion and malathion. Milder, noncovalent (pink) inhibitors of acetylcholinesterase include physostigmine and neostigmine.

Glutamate and Aspartate: Excitatory Amino Acid Neurotransmitters

The common amino acids glutamate and aspartate act as neurotransmitters. Like acetylcholine, glutamate and aspartate are excitatory and stimulate receptors on the postsynaptic membrane to transmit a nerve impulse. The details of glutamate processing at an excitatory synaptic junction are shown in Figure 38.24. In the resting state, the glutamate concentration in the extracellular space is approximately 1 μM, whereas in the presynaptic cytoplasm and the lumen of the synaptic vesicles, the glutamate concentrations are about 10 mM and 100 mM, respectively. These concentrations are maintained by specific carrier proteins. The carriers in the presynaptic and glial plasma membranes are Na$^+$-dependent and have K_m values for glutamate as low as 2 μM. ATPases in the presynaptic vesicle membranes establish a proton gradient, which provides the driving force for accumulation of glutamate anions by a vesicle membrane transporter that has a K_m for glutamate in the millimolar range. Arrival of a nerve impulse triggers Ca^{2+}-dependent exocytosis, in a manner similar to acetylcholine release, followed by binding to glutamate receptors on the postsynaptic membrane. However, since no enzymes that degrade glutamate (in analogy to acetylcholinesterase) exist in the extracellular space, glutamate must be cleared by the high-affinity presynaptic and glial transporters—a process called **reuptake.** Glutamate taken up by glial cells is converted to glutamine by glial glutamine synthetase. Glutamine, which does not display neurotransmitter activity, is present in the extracellular space at about 0.5 mM and is taken into the presynaptic neuron by an Na$^+$-independent carrier. It can be reconverted to glutamate by mitochondrial glutaminase and then accumulated in synaptic vesicles.

There are at least five subclasses of glutamate receptors known. Four of these are identified by specific antagonist effects on glutamate receptors (Figure 38.25). The **N-methyl-D-aspartate (NMDA), kainate,** and **AMPA** receptors are **ligand-gated ion channels.** The **metabotropic** receptors are G protein-mediated receptors that are coupled to phosphatidylinositol metabolism, much in the manner of the muscarinic acetylcholine receptors. The best understood of these excitatory receptors is the NMDA receptor. The NMDA receptor is a ligand-gated channel that, when open, allows Ca^{2+} and Na$^+$ to flow into the cell and K$^+$ to flow out of the cell. A unique property of these ion channels is that they are closed by Mg^{2+} ions in a voltage-dependent fashion.

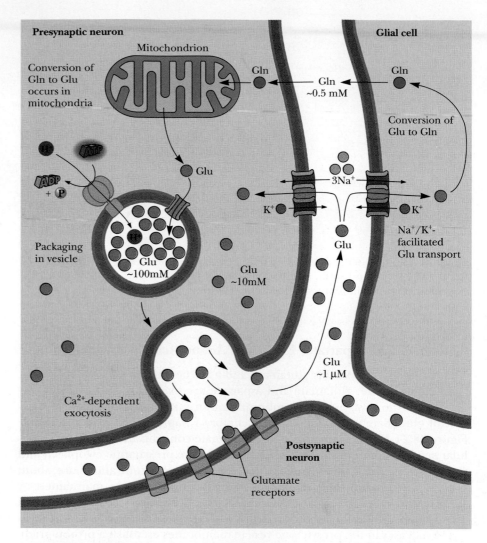

***Figure* 38.24** The excitatory neurotransmitter glutamate is cleared from the synaptic cleft by either of two pathways. K^+-dependent reuptake transporters may transport it back into the presynaptic neuron, or it may accumulate in nearby glial cells via similar transport proteins. Glutamate is repackaged in synaptic vesicles by means of a transport protein that exchanges glutamate for protons. This latter uptake into vesicles depends upon the establishment of a proton gradient across the vesicular membrane. In the glial cells, glutamate is converted to glutamine and then transported back to the presynaptic neuron, where it is converted back to glutamate in the mitochondria.

Phencyclidine (PCP) is a specific antagonist of the NMDA receptor (Figure 38.25), and, unlike Mg^{2+}, its inhibition is not voltage-dependent. Phencyclidine was once used as an anesthetic agent, but legitimate human use was quickly discontinued when it was found to be responsible for bizarre psychotic reactions and behavior in its users. Since this time, PCP has been used illegally as a hallucinogenic drug, under the street name of **angel dust.** Sadly, it has caused many serious, long-term psychological problems in its users.

γ-Aminobutyric Acid and Glycine: Inhibitory Neurotransmitters

Certain neurotransmitters, acting through their conjugate postsynaptic receptors, inhibit the postsynaptic neuron from propagating nerve impulses from other neurons. Two such inhibitory neurotransmitters are **γ-aminobutyric**

(a)

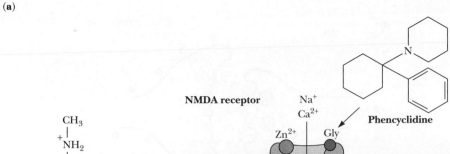
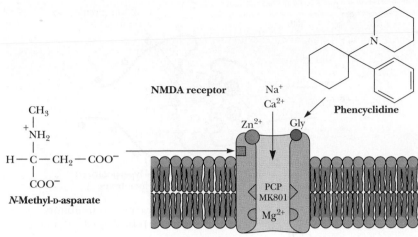

Figure 38.25 Four classes of glutamate receptors. (a) The NMDA receptor is a Na$^+$ and Ca^{2+} channel, which is regulated by Zn^{2+} and glycine, stimulated by *N*-methyl-D-aspartate and inhibited by phencyclidine (PCP) and the anticonvulsant drug MK-801. (b) The kainate receptor is a Na$^+$ and K$^+$ channel. (c) The AMPA receptor is an Na$^+$ channel activated by α-amino-3-hydroxy-5-methyl-4-isoxazolepropionic acid (AMPA), and (d) the metabotropic receptor is a G protein-dependent receptor that is stimulated by ibotenic acid.

(Adapted from Young, A., and Fagg, G., 1990. Excitatory amino acid receptors in the brain: Membrane binding and receptor autoradiographic approaches. Trends in Pharmacological Sciences 11:126–133.)

(b)

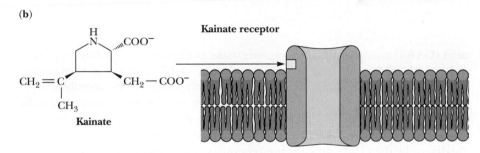

(c)

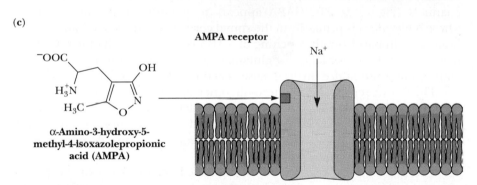

(d)

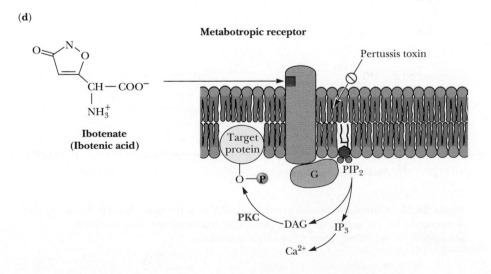

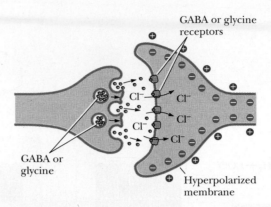

Figure 38.26   GABA (γ-aminobutyric acid) and glycine are inhibitory neurotransmitters that activate chloride channels. Influx of Cl⁻ causes a hyperpolarization of the postsynaptic membrane.

acid (GABA) and **glycine.** These agents make postsynaptic membranes permeable to chloride ions and cause a net influx of Cl⁻, which in turn causes **hyperpolarization** of the postsynaptic membrane (making the membrane potential more negative). Hyperpolarization of a neuron effectively raises the threshold for the onset of action potentials in that neuron, making the neuron resistant to stimulation by excitatory neurotransmitters. These effects are mediated by the GABA and glycine receptors, which are ligand-gated chloride channels (Figure 38.26). GABA appears to operate mainly in the brain, whereas glycine acts primarily in the spinal cord. The effects of ethanol on the brain arise in part from the opening of GABA receptor Cl⁻ channels. GABA is derived by a decarboxylation of glutamate and is broken down by successive aminotransferase and dehydrogenase reactions (Figure 38.27).

The GABA receptor is a hetero-oligomer of 220 to 400 kD, composed of two to four homologous polypeptides. A model has been proposed (Figure 38.28). Positively charged amino acids located at the membrane surfaces, as well as arginine residues within the second transmembrane domain, appear to provide attractive binding sites for Cl⁻.

The glycine receptor can be distinguished on the basis of its specific affinity for the convulsive alkaloid **strychnine** (Figure 38.29). Purified glycine

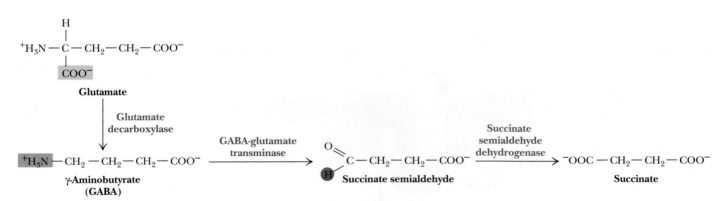

Figure 38.27   Glutamate is converted to GABA by glutamate decarboxylase. GABA is degraded by the action of GABA-glutamate transaminase and succinate semialdehyde dehydrogenase to produce succinate.

(a) GABA$_A$ receptor protein subunit

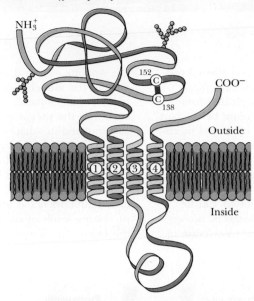

(b) GABA$_A$ receptor-chloride channel protein complex (top view)

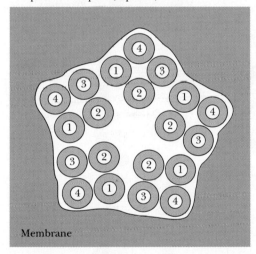

***Figure* 38.28**   (a) A model for the arrangement of the GABA receptor protein subunit in the postsynaptic membrane. (b) Five receptor subunits are postulated to form a transmembrane Cl$^-$ channel. In this model, the walls of the ion channel are formed by the second membrane-spanning segment of each of the five subunits.

(Adapted from Olsen, R., and Tobin, A., 1990. Molecular biology of GABA$_A$ receptors. The FASEB Journal 4:1469–1480.)

receptor from spinal cord tissue is a glycoprotein containing polypeptides of 48 kD, 58 kD, and 93 kD. The 48-kD subunit contains the strychnine-binding site. The 48-kD and 58-kD subunits are homologous integral membrane proteins that are thought to form the Cl$^-$ channel core of the receptor, whereas the 93-kD subunit is a peripheral membrane protein located on the cytoplasmic face of the postsynaptic glycine receptor complex. The sequence of the 48-kD subunit is homologous with that of the GABA receptor and the nicotinic acetylcholine receptor.

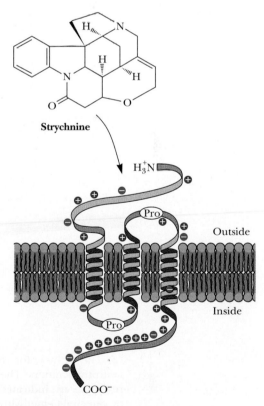

***Figure* 38.29**   Glycine receptors are distinguished by their unique affinity for strychnine. A model for the arrangement of the 48-kD subunit of the glycine receptor in the postsynaptic membrane is shown.

A Deeper Look

The Biochemistry of Neurological Disorders

Defects in catecholamine processing are responsible for the symptoms of many neurological disorders, including clinical depression (which involves norepinephrine) and Parkinsonism (involving dopamine). Norepinephrine (NE) and dopamine (DA) (see Figure 38.30) are processed much the way glutamate is (Figure 38.24). Once these neurotrans-mitters have bound to and elicited responses from postsynaptic membranes, they must be efficiently cleared from the synaptic cleft (see figure below, part a). Clearing can occur by several mechanisms. NE and DA transport or reuptake proteins exist both in the presynaptic membrane and in nearby glial-cell membranes. On the other hand, catecholamine neurotransmitters can be metabolized and inactivated by two enzymes: **catechol-O-methyl-transferase** in the synaptic cleft and **monoamine oxidase** in the mitochondria (see figure part b). Catecholamines transported back into the presynaptic neuron are accumulated in synaptic vesicles by the same H^+-ATPase/H^+-ligand exchange mechanism described for glutamate. Clinical depression has been treated by two different strategies. **Monoamine oxidase inhibitors** act as antidepressants by increasing levels of catecholamines in the brain. Another class of antidepres-

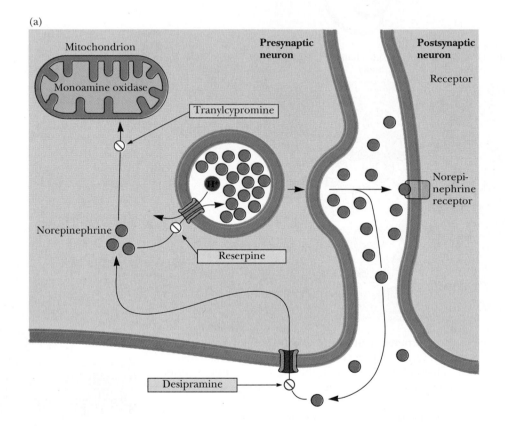

(a) The pathway for reuptake and vesicular repackaging of the catecholamine neurotransmitters. The sites of action of desipramine, tranylcypromine, and reserpine are indicated. (b) Norepinephrine can be degraded in the synaptic cleft by catechol-O-methyltransferase or in the mitochondria of presynaptic neurons by monoamine oxidase. (c) The structures of tranylcypromine and desipramine. (d) The structure of bromocriptine. (e) The structure of reserpine. (f) The structure of cocaine.

sants, the **tricyclics,** such as desipramine (see figure part c) act specifically on NE reuptake transporters and facilitate more prolonged stimulation of postsynaptic NE receptors.

Parkinsonism is characterized by degeneration of dopaminergic neurons, as well as consequent overproduction of postsynaptic dopamine receptors. In recent years, Parkinson's patients have been treated with dopamine agonists such as bromocriptine (see figure part d) to counter the degeneration of dopamine neurons.

Catecholaminergic neurons are involved in many other interesting pharmacological phenomena. For example, **reserpine** (see figure part e), an alkaloid from a climbing shrub of India, is a powerful sedative that depletes the level of brain monoamines by inhibiting the H^+-monoamine exchange protein in the membranes of synaptic vesicles.

Cocaine (see figure part f), a highly addictive drug, binds with high affinity and specificity to reuptake transporters for the monoamine neurotransmitters in presynaptic membranes. Thus, at least one of the pharmacological effects of cocaine is to prolong the synaptic effects of these neurotransmitters.

(b)

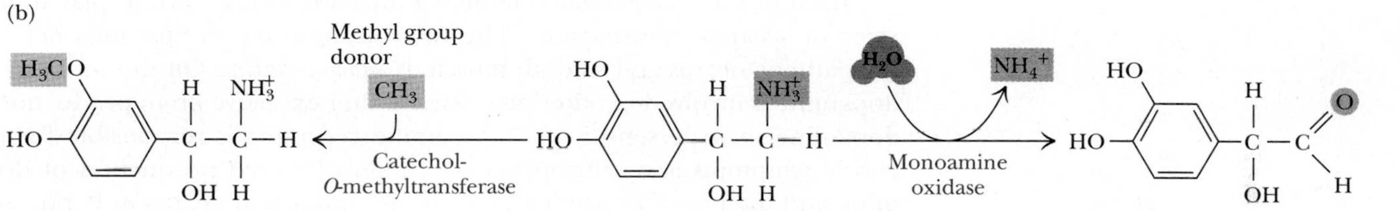

3-O-Methylepinephrine   **Norepinephrine**   **3,4-Dihydroxyphenylglycolaldehyde**

(c)

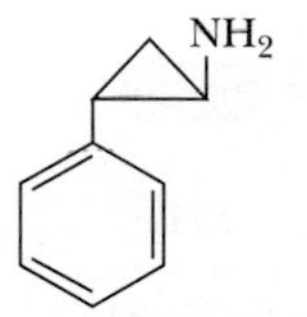

Tranylcypromine 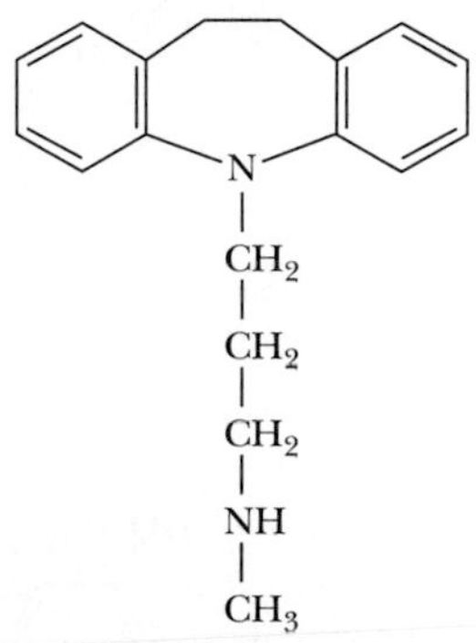   **Desipramine**

(d)

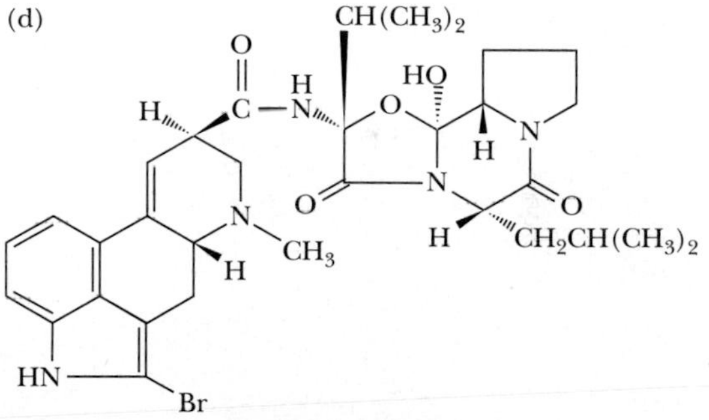

Bromocriptine

(e)

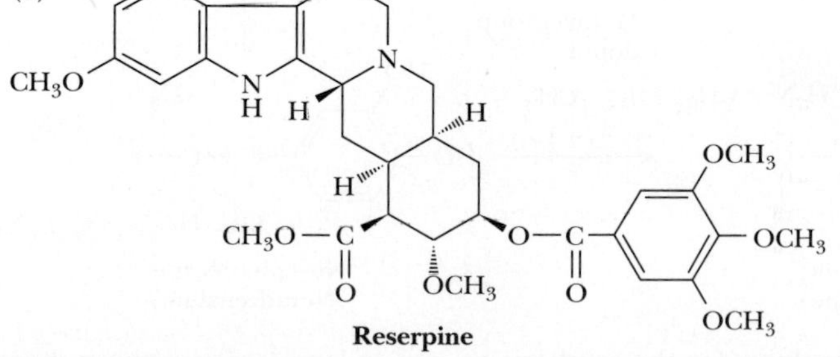

Reserpine

(f)

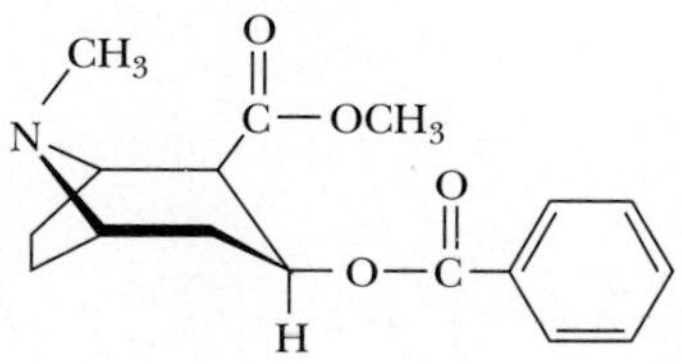

Cocaine

The Catecholamine Neurotransmitters

Epinephrine, norepinephrine, dopamine, and **L-dopa** are collectively known as the **catecholamine** neurotransmitters. These compounds are synthesized from tyrosine (Figure 38.30), both in sympathetic neurons and in the adrenal glands. They function as neurotransmitters in the brain and as hormones in the circulatory system. However, these two pools operate independently, thanks to the **blood–brain barrier,** which permits only very hydrophobic species in the circulatory system to cross over into the brain. [In spite of this, the *brain receptors* are referred to as **adrenergic** receptors (Chapter 37).] Hydroxylation of tyrosine (by **tyrosine hydroxylase**) to form **3,4-*dihydroxyphenylalanine*** (L-dopa) is the rate-limiting step in this pathway. Dopamine, a crucial catecholamine involved in several neurological diseases, is synthesized from L-dopa by a pyridoxal phosphate-dependent enzyme, **dopa decarboxylase.** Subsequent hydroxylation and methylation produce norepinephrine and epinephrine (Figure 38.30). The methyl group in the final reaction is supplied by S-adenosylmethionine.

Each of these catecholamine neurotransmitters is known to play unique roles in synaptic transmission. The neurotransmitter in junctions between sympathetic nerves and smooth muscle is *norepinephrine.* On the other hand, dopamine is involved in other processes. Either excessive brain production of dopamine or hypersensitivity of dopamine receptors is responsible for psychotic symptoms and schizophrenia, whereas lowered production of dopamine and the loss of dopamine neurons are important factors in Parkinson's disease.

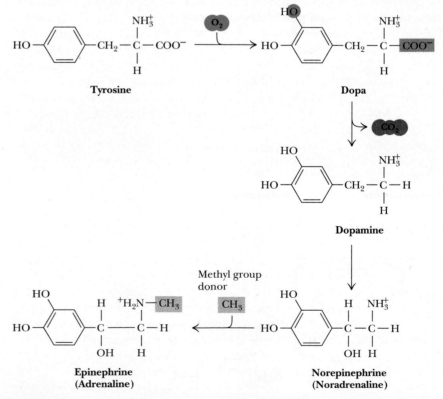

***Figure* 38.30** The pathway for the synthesis of catecholamine neurotransmitters. Dopa, dopamine, noradrenaline, and adrenaline are synthesized sequentially from tyrosine.

At least three different dopamine receptors (denoted D_1, D_2, and D_3) have been characterized. These dopamine receptor subtypes are homologous with one another, as well as with the β-adrenergic receptors, and are putative 7-TMS proteins of 446 residues. All possess the conserved Asp residue in the third transmembrane segment, which is found in the β-adrenergic receptor sequences. In addition, they contain two Ser residues in the fifth transmembrane segment that are conserved among catecholamine receptors and are critical for the recognition of agonist ligands possessing a catechol group. **D_1 receptors** stimulate adenylyl cyclase. **D_2 receptors** have been linked to inhibition of adenylyl cyclase, inhibition of phosphatidylinositol turnover, activation of K^+ channels, and inhibition of Ca^{2+}-channel activity. **D_3 receptors** are similar to D_2 receptors.

The Peptide Neurotransmitters

Many relatively small peptides have been shown to possess neurotransmitter activity (see Table 38.1). One of the challenges of this field is that the known neuropeptides may represent a very small subset of the neuropeptides that exist. Another challenge arises from the small *in vivo* concentrations of these agents, and the small number of receptors that are present in neural tissue. Physiological roles for most of these peptides are complex. For example, the **endorphins** and **enkephalins** are natural opioid substances and potent pain relievers. The **endothelins** are a family of homologous regulatory peptides, synthesized by certain endothelial and epithelial cells, that act on nearby smooth muscle and connective tissue cells. They induce or affect smooth muscle contraction; vasoconstriction; heart, lung, and kidney function; as well as mitogenesis and tissue remodeling. **Vasoactive intestinal peptide (VIP)** produces a G–protein-adenylyl cyclase-mediated increase in cAMP, which in turn triggers a variety of protein phosphorylation cascades, one of which leads to conversion of phosphorylase *b* to phosphorylase *a*, stimulating glycogenolysis. Moreover, VIP has synergistic effects with other neurotransmitters, such as norepinephrine. In addition to increasing cAMP levels through β-adrenergic receptors, norepinephrine acting at α_1-adrenergic receptors markedly stimulates the increases in cAMP elicited by VIP. Many other effects have also been observed. For example, injection of VIP increases rapid eye movement (REM) sleep and decreases waking time in rats. How these complex effects are mediated is not understood, but it has been shown that VIP receptors exist in regions of the central nervous system involved in sleep modulation.

38.7 Sensory Transduction

The process of sensing the environment and turning the acquired information into nerve impulses is termed **sensory transduction.** Light and images, sounds, tastes, smells, and tactile sensations (shapes, textures, movements, heat and cold) must all be translated into neural signals, and most organisms must also be able to sense pain. More complex sensations, such as fear, anger, and sexual attraction, to name a few, are complex combinations of the simple ones.

All of these systems have certain features in common. In all cases, highly specialized sensory cells translate the sensory stimulus into an electrochemical potential change, which is communicated to adjacent neurons. The intensity of a stimulus is determined by the number and frequency of action potentials produced by that sensory system.

Vision

The fundamental event that makes the visual system work is the absorption of light quanta by **rhodopsin,** which is formed from the protein **opsin** and a molecule of **11-*cis*-retinal.** Light absorption causes isomerization of the retinal, which initiates a sequence of molecular events resulting in the stimulation of an action potential in an adjacent neuron. The anatomy of the eye and of the retina are shown in Figure 38.31.

Rod and cone cells lie at the base of several layers of cells in the retina, with their light-sensitive portions at the extreme rear of this structure. Light must pass through these cells in order to reach the rhodopsin molecules themselves. Rods and cones are structurally elaborate cells approximately 40 μm in length and about 1 μm in diameter (Figure 38.32). The so-called **inner segment** contains a synaptic body (for connection with bipolar neurons). The **outer segments,** where photons are absorbed, are attached to the inner segment by means of a thin, ciliary structure. The outer segment contains a stack of approximately 2000 **disks,** which are flattened vesicles derived from the

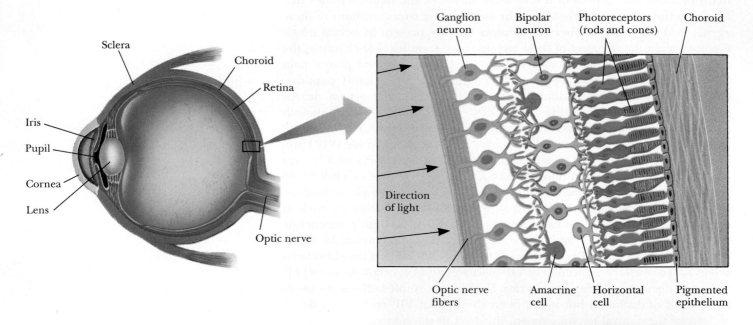

***Figure* 38.31** The anatomy of the human eye and a cross-section of the human retina. Light falling on the eye passes through the cornea and is focused by the lens on the retina, a complex neural structure that contains both rod cells and cone cells (named for their approximate shapes). The typical human retina has about 100 million rod cells and only about 3 million cones. Rods are more sensitive and function well at low light levels, but they are not capable of distinguishing colors. Cones distinguish colors but are less sensitive to light. There is **two-dimensional organization** on the surface of the retina. The **macula,** the portion of the retina that corresponds to the center of the field of vision, is rich in cones and contains relatively fewer rod cells. The surrounding area, which is responsible for peripheral vision, is poor in cones but richer in rods. The cone-rich macula provides the visual acuity needed for reading, but it is a relatively insensitive detector of low light levels.

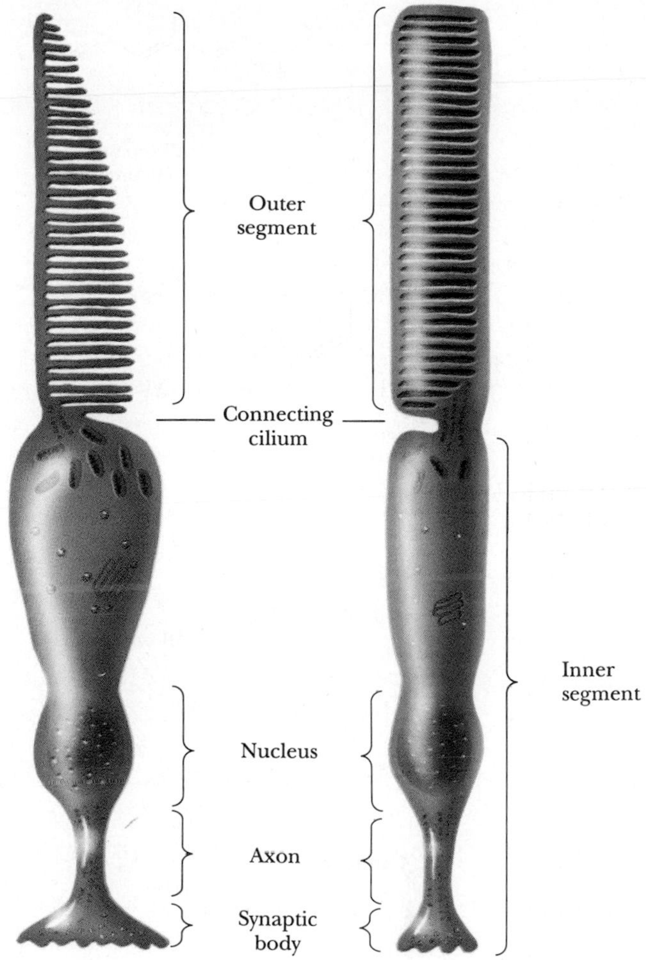
Outer segment
Connecting cilium
Inner segment
Nucleus
Axon
Synaptic body

***Figure* 38.32** Schematic drawing of cone and rod cells. Rod outer segment (ROS) disks are continually formed by invaginations of the plasma membrane near the middle of the cell. At the back of the rod cell, old disks are expelled and degraded by epithelial cells at the base of the retina. This pair of processes gives the average disk a life span of about 7 to 10 days. In cone cells, the invaginations never close off to form disks and the rhodopsin-containing membranes are continuous with the plasma membrane. Regeneration of cone cell membrane components occurs continuously throughout the length of the outer segment.

plasma membrane. The disk membranes are densely packed with rhodopsin. Each disk contains approximately 15,000 rhodopsin molecules, and a single rod cell thus contains about 30 million molecules of rhodopsin.

Rhodopsin in the rod outer segment (ROS) disks is an integral 40-kD protein with seven transmembrane α-helices (Figure 38.33). The amino terminal segment, which lies inside the disks, contains two N-linked glycosylation sites, whereas the C-terminus, which faces the cytoplasm, contains multiple regulatory phosphorylation sites (Ser and Thr).

Opsin by itself does not absorb visible light, but rhodopsin does, owing to the presence of its retinal chromophore. The precursors for retinal cannot be synthesized by mammals, but retinal can be derived from retinol (vitamin A) or β-carotenes acquired in the diet (Figure 38.34). The 11-*cis* isomer of retinal has the absorption spectrum shown in Figure 38.35, with a maximum at about 380 nm. When 11-*cis*-retinal is bound to rhodopsin, however, its absorption maximum shifts to approximately 500 nm. Moreover, the *intensity* of the absorption increases as well, reaching molar absorptivity at 500 nm of just over

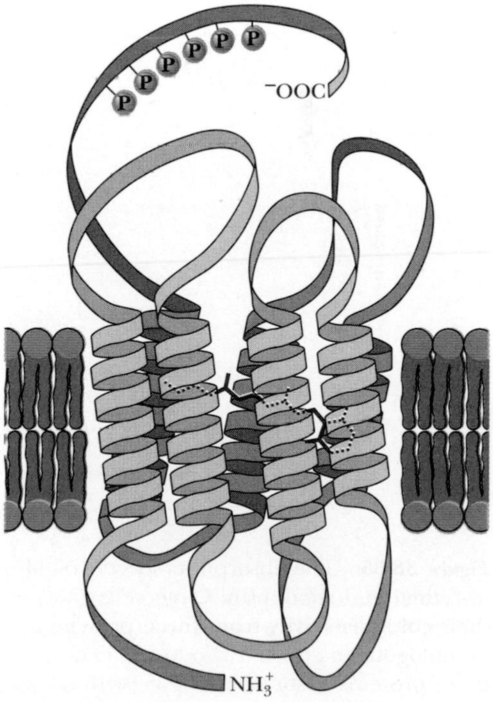
P
P
P
P
P
P
P
$^{-}$OOC
NH_3^+

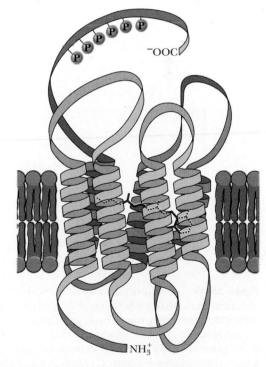

***Figure* 38.33** A model for rhodopsin, the photon receptor of retinal rod outer segment membranes. Note the C-terminal phosphorylation sites.

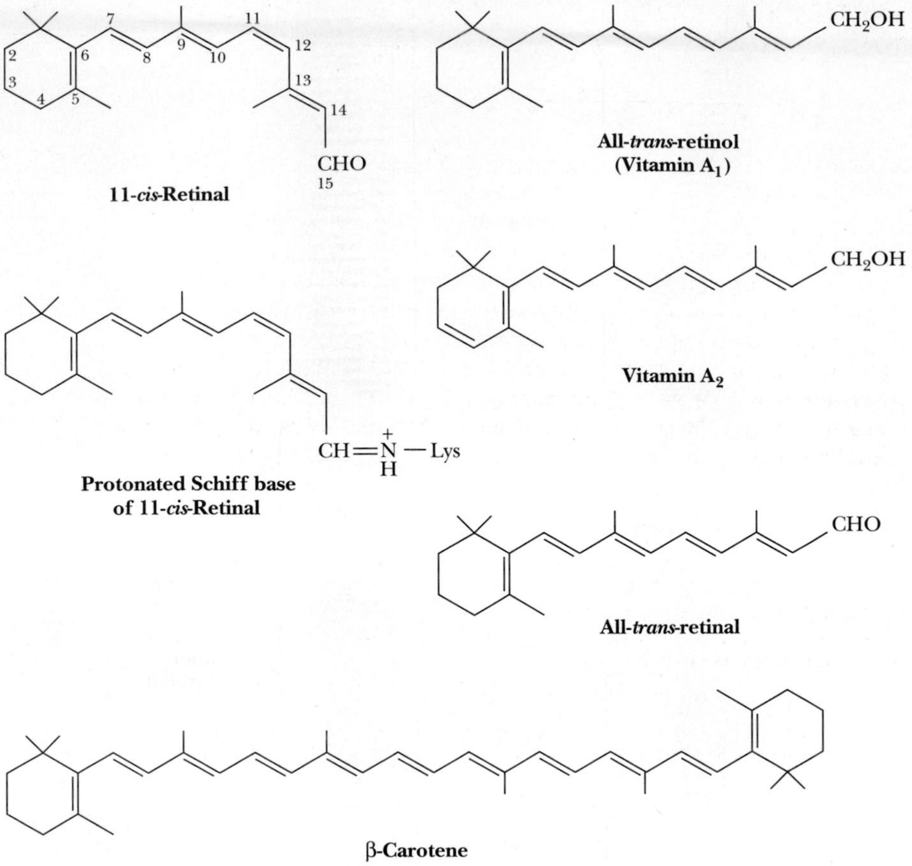

11-*cis*-Retinal

**All-*trans*-retinol
(Vitamin A₁)**

Vitamin A₂

**Protonated Schiff base
of 11-*cis*-Retinal**

All-*trans*-retinal

β-Carotene

***Figure* 38.34**   Structures of 11-*cis*-retinal, all-*trans*-retinal, vitamins A₁ and A₂, and β-carotene.

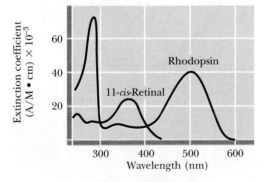

***Figure* 38.35**   The absorption spectra of 11-*cis*-retinal and rhodopsin. Cone cells gain their color sensitivity from three proteins homologous to rod cell rhodopsin. These three proteins absorb blue light (with λ_{max} = 440 nm), green light (λ_{max} = 530 nm), and yellow light (λ_{max} = 570 nm) and pass separate signals to the brain for processing.

40,000 $M^{-1}cm^{-1}$, a very high value. The efficient absorption of light by retinal is due to its highly conjugated structure and its favorable interactions with opsin.

11-*cis*-Retinal is bound to rhodopsin in a deep cleft; it lies nearly parallel to the membrane sheet, near the middle of the membrane, and is surrounded by the transmembrane segments of the protein. The aldehyde function of retinal forms an important Schiff base linkage with Lys[296] in the seventh transmembrane helix (Figure 38.33). Absorption of a photon of light by rhodopsin causes *the isomerization of 11-cis-retinal to the all-trans isomer*. This change involves several rapidly formed intermediate states (Figure 38.36). The intermediate states were first isolated and characterized by illuminating rhodopsin under liquid nitrogen (b.p. 77K), and then slowly raising the temperature to allow transitions between intermediate states. Absorption of a photon at 77K immediately shifts the λ_{max} from 500 nm to 543 nm—a state called **bathorhodopsin.** If this state is warmed to more than 130K, it rapidly converts to **lumirhodopsin** (λ_{max} = 497 nm). Warming this state above 230K yields **metarhodopsin I** (λ_{max} = 478 nm), and above 255K, **metarhodopsin II,** often denoted **R*** (λ_{max} = 380 nm), is formed. In this state, the crucial Schiff base is deprotonated. Above 273K, the retinal, now in the all-*trans* configuration, dissociates from rhodopsin. Isomerization of all-*trans*-retinal back to 11-*cis*-retinal facilitates the binding of retinal to opsin and regenerates rhodopsin. The essential features of this model have been confirmed in recent years by rapid kinetic measurements at physiological temperature.

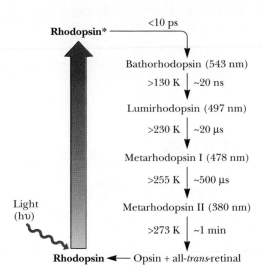

Figure 38.36 Following excitation of rhodopsin by a photon of light, intermediate states are formed in the sequence shown. The optical absorption maxima are indicated in parentheses for each state. NMR and resonance Raman spectroscopy data show that, in the resting (dark) state, the Schiff base linking retinal to the protein is protonated.

A crucial feature of the metarhodopsin II state is the exposure or unmasking of cytoplasmic domains of rhodopsin. This allows metarhodopsin (R*) to interact with other proteins in the ROS. This initiates a series of reactions that lead to a change in cation permeability of the plasma membrane, which in turn initiates a nerve impulse. The essential features of this process are shown in Figure 38.37. Na^+,K^+-ATPase in the inner segment membrane establishes and maintains gradients of sodium and potassium. Potassium leakage out of the cell establishes a potential gradient of approximately 20 mV (negative inside) across the inner segment membrane. Na^+, on the other hand, does not easily leak back into the inner segment. Instead, it reenters the rod cell through specific channels in the ROS membrane, as shown. In this way, the resting (dark) rod cell maintains a steady-state balance between cation pumping and leakage. However, stimulation of the rod cell by light causes many of the Na^+ channels in the ROS to close quickly. This quick closing causes the electrical potential of the rod cell to become more negative—a state of hyperpolarization. The result is release of a neurotransmitter substance into the synaptic cleft between the rod cell and a bipolar cell, triggering a nerve impulse to the brain.

Rod cells are among the most sensitive of biological devices. *The absorption of a single photon of light by a single rhodopsin molecule can trigger a nerve impulse by that rod cell.* How is it that a single light-absorption event can be transduced and amplified to produce the closure of many Na^+ channels? The answer involves a specific GTP-binding protein and a cyclic nucleotide second messenger, cGMP. As shown in Figure 38.38, activated rhodopsin (R*) induces a GTP–GDP exchange by **transducin,** a typical G protein, which in turn stimulates a **phosphodiesterase (PDE)** to hydrolyze cGMP to GMP. The plasma membrane Na^+ channels, which are gated open by cGMP, rapidly close when cGMP levels decrease. The needed amplification in this **cascade** of reactions is provided by R*, which activates hundreds of transducins, and the phosphodiesterase, which hydrolyzes thousands of cGMP molecules per second.

Transducin is a peripheral membrane protein located on the cytoplasmic surface of the ROS disks. Like other G proteins, transducin ($T_{\alpha\beta\gamma}$) is an 80-kD

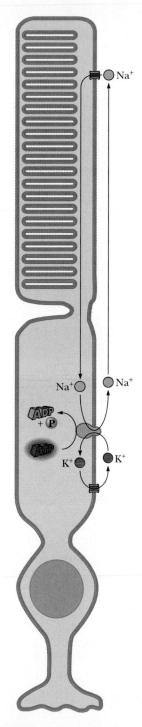

Figure 38.37 In the resting state, rod cells maintain a steady-state level of Na^+ ions that is determined by the actions of Na^+, K^+-ATPase in the inner segment, and Na^+ channels in the outer segment. The Na^+ channels are gated open by cGMP. Excitation of rhodopsin by light leads to a reduction in cGMP concentration, which results in the closing of the Na^+ channels.

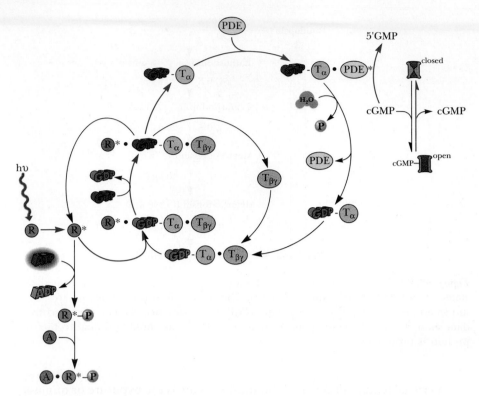

***Figure* 38.38** Excitation of rhodopsin (R) activates transducin ($T_{\alpha\beta\gamma}$), a G protein, which exchanges GDP for GTP. The T_α(GTP) complex dissociates from $T_{\beta\gamma}$ and binds to phosphodiesterase, which converts cGMP to 5′-GMP. The GTPase activity of T_α converts bound GTP to GDP, and the T_α(GDP) complex dissociates from phosphodiesterase. Reassociation of T_α(GDP) with $T_{\beta\gamma}$ returns transducin to the resting state. Red arrows indicate the activation steps of the visual cascade. *A* stands for arrestin (see page 1250).

(Adapted from Lolley, R., and Lee, R., 1990. Cyclic GMP and photoreceptor function. The FASEB Journal 4:3001–3008.)

heterotrimer composed of α- (39 kD), β- (37 kD), and γ- (8.5 kD) subunits. T_α possesses a guanosine nucleotide-binding site. The T_α and T_γ subunits are unique to rod cells, but T_β is identical to β-subunits found in other G proteins. In the dark resting state, GDP is bound to T_α, which remains tightly complexed with the β- and γ-subunits, and the $T_{\alpha\beta\gamma}$ complex remains loosely associated with the disk membrane. However, light absorption changes the conformation of rhodopsin, exposing a surface in R* that binds T_α. The R* T_α complex rapidly releases GDP and binds GTP. This reaction induces a conformation change in T_α that causes T_α(GTP) to dissociate from R* and from $T_{\beta\gamma}$, (Figure 38.39). Binding of T_α(GTP) to the cytoplasmic phosphodiesterase then activates hydrolysis of cGMP.

As indicated above, a key to the amplification that occurs in visual transduction is that one R* can activate hundreds of molecules of transducin. The ROS disk membrane is a highly fluid structure. R* diffuses freely in the plane of the lipid bilayer (with a lateral diffusion coefficient of 0.5 μm^2/sec). This free lateral motion allows R* to interact with and activate approximately 500 molecules of transducin during its activated lifetime.

The phosphodiesterase complex of rod outer segments, which is activated by the binding of T_α(GTP), is a heterotetramer consisting of one α-subunit (88 kD), one β-subunit (84 kD), and two small γ-subunits (11 kD each). The α- and β-subunits are tightly associated and both may have catalytic activity. The γ-subunits inhibit phosphodiesterase activity by α and β. T_α(GTP) activates this enzyme by binding to a γ-subunit and dissociating it from the $\alpha\beta\gamma_2$

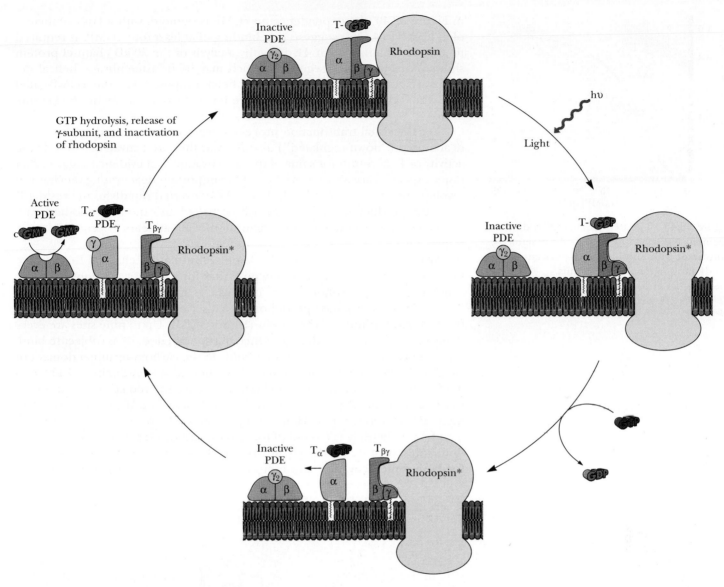

Figure 38.39 $T_\alpha(GTP)$ activates phosphodiesterase by facilitating the dissociation of γ subunits from the phosphodiesterase $\alpha\beta\gamma_2$ complex.

complex. Removal of one γ-subunit activates the phosphodiesterase to half its maximal activity (Figure 38.39). If another $T_\alpha(GTP)$ should bind to the second γ-subunit, this also is displaced, rendering the $\alpha\beta$ complex fully active. This phosphodiesterase is highly efficient; it has a turnover number of more than $4 \times 10^3 \, s^{-1}$ and a k_{cat}/K_m of approximately $6 \times 10^7 \, M^{-1}s^{-1}$—a value near the theoretical diffusion limit for bimolecular reactions. The phosphodiesterase remains active as long as the γ-subunit remains dissociated and complexed to $T_\alpha(GTP)$. However, T_α itself is a GTPase, and slowly hydrolyzes the GTP that activates it. Hydrolysis of GTP to GDP on T_α releases the phosphodiesterase γ-subunit, which can then bind and inactivate the phosphodiesterase. By the time this has occurred, however, activation of a single rhodopsin has caused the hydrolysis of approximately 100,000 molecules of cGMP.

As noted above, the decrease in cGMP concentration by phosphodiesterase leads to hyperpolarization of the rod cell membrane by closing ion channels. These ion channels allow the passive influx of Na^+ and Ca^{2+} under dark conditions, with Na^+ flux accounting for 85% of the dark current and Ca^{2+} for 15%. These ligand-gated ion channels are modulated by cGMP. As shown

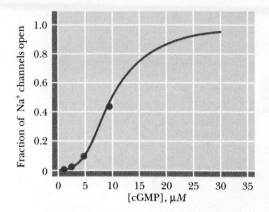

***Figure* 38.40**   The dependence of rod outer segment Na⁺ channels on cGMP is sigmoid, consistent with an oligomeric association of cGMP-binding channel subunits.

(Data from Zimmerman, A. L., and Baylor, D. A., 1986. Cyclic GMP-sensitive conductance of retinal rods consists of aqueous pores. Nature 321:70–72.)

in Figure 38.40, the dependence on cGMP is sigmoid, with a Hill coefficient of at least 3 to 4. This suggests that binding of at least four cGMPs is required to gate the channel open. Hydropathy analysis of the 80-kD channel protein shows six hydrophobic segments, which may be transmembrane helical domains (Figure 38.41). A model has been proposed for the cGMP-gated Na⁺/Ca²⁺ channel that involves a pore formed from four or five 63-kD subunits.

For the visual transduction process to work effectively, rod cells must be deactivated ("down-regulated") as rapidly as they are activated. The GTPase activity of T_α accounts for some of this deactivation, but evidence suggests that faster processes are also at work here. One important deactivating mechanism involves the quenching of R*. If R* can be converted rapidly to the resting R state, the production of T_α(GTP) will be halted and the phosphodiesterase will be quickly inactivated. The inactivation of R* appears to proceed in two steps.

When R is activated to R*, the same conformation changes that expose binding sites for transducin also expose sites for a specific **rhodopsin kinase,** which begins to phosphorylate R* even as it is activating transducins (Figure 38.42). Rhodopsin kinase phosphorylates a series of Ser and Thr residues at the carboxyl terminus of rhodopsin (Figure 38.33). Up to nine sites are eventually phosphorylated in this way. Once phosphorylated, R* is subject to binding by **arrestin,** a 48-kD protein that binds to the carboxy-terminal domain of rhodopsin, competitively inhibiting interaction with transducin and preventing further activation of phosphodiesterase. Since the rod cell contains much more transducin (500 μM) than rhodopsin kinase (5 μM), transducin binding to R* is favored at first. However, upon binding arrestin, each phosphorylated R* is effectively taken out of the action, and rhodopsin kinase eventually dominates, turning off the visual cascade. This sequence of reactions allows for both rapid activation and rapid deactivation in visual transduction.

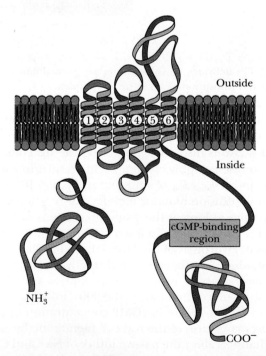

***Figure* 38.41**   A model for the arrangement of cGMP-dependent Na⁺ channel subunits in the rod outer segment membrane.

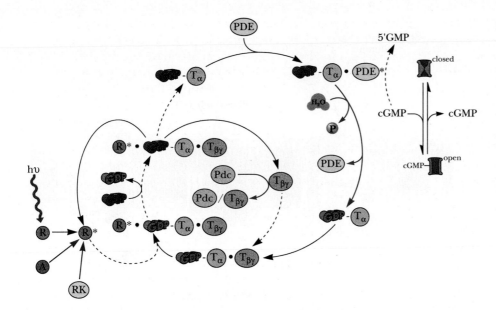

***Figure* 38.42** Red arrows indicate the reactions that quench the phototransduction cascade, as does the intrinsic GTPase activity of T_α, and the binding of $T_{\beta\gamma}$ by phosducin, or the binding of R* by arrestin. Phosphorylation of activated rhodopsin by rhodopsin kinase (RK) leads to quenching of the phototransduction cascade.

(Adapted from Lolley, R., and Lee, R., 1990. Cyclic GMP and photoreceptor function. The FASEB Journal 4:3001–3008.)

Olfaction

The biochemistry of olfaction—the sense of smell—has much in common with the biochemistry of other excitable membranes and hormonally regulated systems we have already considered. In mammals, odorant molecules—those that have a smell—are carried (by breathing) up the nasal passages to the **primary olfactory epithelium** (Figure 38.43). Here they diffuse through a layer of mucus and interact with cilia projecting into the mucus from the **olfactory receptor neurons.** The binding of odorants to as-yet-unidentified receptors makes the transmembrane potential of the receptor neurons less negative. This depolarization increases the rate at which action potentials are sent down the receptor neuron axons to the olfactory bulb in the brain. The mechanism is believed to involve activation of a unique olfactory G protein, G_{olf}, by the odorant receptors on the cilia. The α-subunit of G_{olf} is a 44-kD protein that shows 88% amino acid identity with G_s (Chapter 37). G_{olf} activates a unique adenylyl cyclase, found only in olfactory neurons and referred to as a **type III adenylyl cyclase,** which produces cAMP in the receptor neurons. Binding of cAMP to cAMP-gated ion channels causes these channels to open, allowing Na^+ ions to flow into the cell, depolarizing the cell and increasing the rate of action potential firing. However, some olfactory responses occur by mechanisms that do not involve cAMP. Certain odorants have been shown to cause a rapid rise in inositol trisphosphate (IP₃) in olfactory receptor neurons. In addition, Ca^{2+} channels in these receptor neurons are opened by IP₃. Thus, as shown in Figure 38.44, at least two different processes appear to be triggered by odorant binding. No odorant receptor proteins have yet been identified or characterized. There is assumed to be an array of receptor molecules, each capable of binding different odorants, but it is not known how broad this array might be or how information about odors is coded by receptor cells for transmission to the brain.

Hearing

Among the senses, hearing and touch are unique in that they transduce mechanical stimuli, as opposed to light or chemical stimuli. Sound waves entering the ear create an oscillating air pressure on the eardrum. The bones in the middle ear essentially convert these airborne vibrations into an oscillating

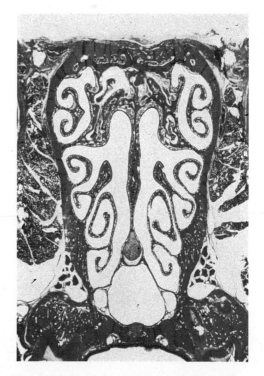

***Figure* 38.43** A cross-section of the nasal passages of a rat. The olfactory epithelium is a layer of receptor cells (and supporting cells) surrounding the air spaces. The spiral structures, called turbinates, increase the surface area of olfactory epithelium exposed to airborne odorant molecules.

(From Taylor, R., 1991. Whiff and poof: adenylyl cyclase in olfaction. The Journal of NIH Research 3:49–53. Photo courtesy of Randall Reed, Johns Hopkins University.)

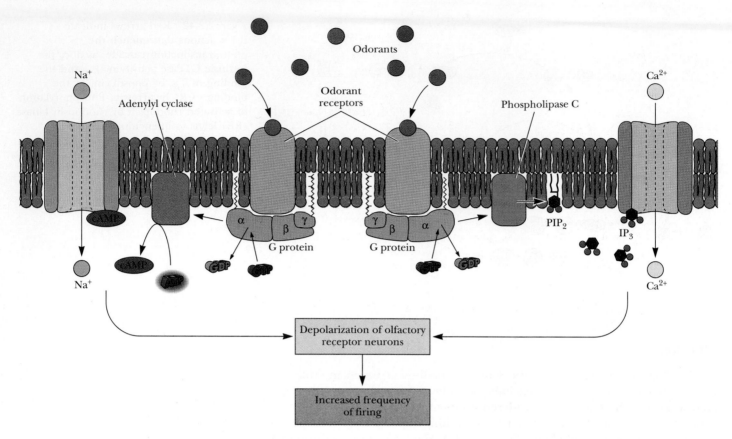

***Figure* 38.44**   A putative signal transduction pathway in olfactory receptor neurons.

pressure in the fluids of the human inner ear, in which are immersed an array of some 16,000 **hair cells.** The "hairs" extending from these cells are actually cross-linked bundles of actin filaments.

Transduction of sound waves into electrical signals results from movements or deflections of the actin filaments. Deflections of the filaments in one direction cause depolarization of the hair cell, whereas deflections toward the other end cause hyperpolarization. The physical nature of the cation channels involved in transduction is not understood. However, a few of the details of the ear's ability to distinguish different frequencies of sounds are known. There are three known mechanisms: (a) different hair cells are mechanically sensitive to different frequencies, (b) the length and mechanical properties of a given bundle of actin filaments make it sensitive to a limited range of frequencies, and (c) the particular balance of ion channel response characteristics in a given hair cell effectively "tunes" the cell. As shown in Figure 38.45, this last mechanism involves an interplay between the transduction channels for K^+, voltage-gated Ca^{2+} channels, and Ca^{2+}-activated K^+ channels. A positive deflection of a hair bundle allows K^+, the principal cation in the fluid that surrounds the bundles, to enter through the transduction channels and cause depolarization of the cell. This depolarization activates the voltage-sensitive Ca^{2+} channels, and the influx of Ca^{2+} causes further depolarization of the cell. Accumulated Ca^{2+}, however, opens the Ca^{2+}-sensitive K^+ channels, and the efflux of K^+ through these channels repolarizes the membrane and the next cycle of the oscillation ensues. A single deflection of the hair bundle results in a damped sinusoidal oscillation of the membrane potential. *The frequency of this sinusoidal oscillation is identical with the acoustically induced fre-*

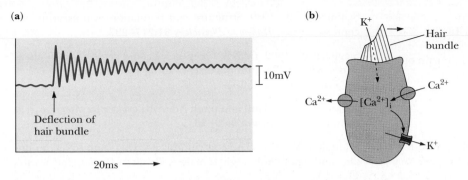

(a)

Deflection of hair bundle

10mV

20ms ———→

(b)

K⁺ → Hair bundle

Ca²⁺

Ca²⁺ ← [Ca²⁺]ᵢ ← Ca²⁺

K⁺

***Figure* 38.45** (a) When an auditory hair cell is deflected, the cell membrane potential undergoes damped, sinusoidal oscillations. (b) These oscillations are effected by synchronized opening and closing of Ca^{2+} and K^+ channels. The frequency of the oscillations matches the optimum response frequency of the hairs on the hair cell.

(Adapted from Roberts, W., Howard, J., and Hudspeth, A., 1988. Hair cells: transduction, tuning and transmission in the inner ear. Annual Review of Cell Biology 4:63–92.)

quency to which the cell is most sensitive. Thus, the electrical properties of the hair cell enable it to **resonate,** drawing energy from the ion gradients to enhance the acoustically delivered oscillations.

Problems

1. Use the Nernst equation to calculate the equilibrium potential across a membrane that separates a 330 mM Na⁺ solution (side 1) from a 70 mM Na⁺ solution (side 2).

2. Use the Goldman equation to calculate the actual potential difference at steady state across a membrane that separates a mixture of 30 mM Na⁺, 70 mM K⁺, and 80 mM Cl⁻ (side 1) from a solution of 100 mM Na⁺, 20 mM K⁺, and 140 mM Cl⁻ (side 2).

3. Synaptic vesicles are approximately 40 nm in outside diameter, and each vesicle contains about 10,000 acetylcholine molecules. Calculate the concentration of acetylcholine in a synaptic vesicle.

4. Decamethonium and succinylcholine (Figure 38.22) both inhibit acetylcholinesterase, but decamethonium's effects are much longer-lasting. Explain. One of these is used as an anesthetic in certain surgeries and other medical treatments. Which do you think it is and why?

5. GTPγS is a nonhydrolyzable analog of GTP. Experiments with squid giant synapses reveal that injection of GTPγS into the presynaptic end (terminal) of the neuron inhibits neurotransmitter release (slowly and irreversibly). The calcium signals produced by presynaptic action potentials and the number of synaptic vesicles docking on the presynaptic membrane are unchanged by GTPγS. Propose a model for neurotransmitter release that accounts for all of these observations.

6. When retinal rod cells are illuminated by light, would you expect any change in the plasma membrane potential? If so, what and why?

Further Reading

Barchi, R. L., 1988. Probing the molecular structure of the voltage-dependent sodium channel. *Annual Review of Neurosciences* **11**:455–495.

Barinaga, M., 1991. How the nose knows: Olfactory receptor cloned. *Science* **252**:12–13.

Catterall, W. A., 1988. Structure and function of voltage-sensitive ion channels. *Science* **252**:50–61.

Chuang, D.-M., 1989. Neurotransmitter receptors and phosphoinositide turnover. *Annual Review of Pharamacology and Toxicology* **29**:71–110.

Corwin, J., and Warchol, M., 1991. Auditory hair cells: Structure, function, development, and regeneration. *Annual Review of Neurosciences* **14**:301–333.

Hulme, E., Birdsall, N., and Buckley, N., 1990. Muscarinic receptor subtypes. *Annual Review of Pharmacology and Toxicology* **30**:633–673.

Jackson, H., and Parks, T., 1989. Spider toxins: Recent applications in neurobiology. *Annual Review of Neurosciences* **12**:405–414.

Jones, D., and Reed, R., 1989. G$_{olf}$: An olfactory neuron specific-G protein involved in odorant signal transduction. *Science* **244**:790–795.

Monaghan, D., Bridges, R., and Cotman, C., 1989. The excitatory amino acid receptors: Their classes, pharmacology, and distinct properties in the function of the central nervous system. *Annual Review of Pharmacology and Toxicology* **29**:365–402.

Montal, M., 1990. Molecular anatomy and molecular design of channel proteins. *The FASEB Journal* **4**:2623–2635.

Nathanson, N., 1987. Molecular properties of the muscarinic acetylcholine receptor. *Annual Review of Neurosciences* **10:**195–236.

Petrou, S., et al., 1993. A putative fatty acid-binding domain of the NMDA receptor. *Trends in Biochemical Sciences* **18:**41–42.

Plotkin, M., 1993. *Tales of a Shaman's Apprentice.* New York: Viking Penguin.

Roberts, W., Howard, J., and Hudspeth, A., 1988. Hair cells: Transduction, tuning and transmission in the inner ear. *Annual Review of Cell Biology* **4:**63–92.

Schimerlik, M., 1989. Structure and regulation of muscarinic receptors. *Annual Review of Physiology* **51:**217–227.

Sieghart, W., 1992. GABA$_A$ receptors: Ligand-gated Cl$^-$ ion channels modulated by multiple drug-binding sites. *Trends in Physiological Sciences* **13:**446–450.

Taylor, R., 1991. Whiff and poof: Adenylyl cyclase in olfaction. *The Journal of NIH Research* **3:**49–53.

Whittaker, V., 1990. The contribution of drugs and toxins to understanding of cholinergic function. *Trends in Physiological Sciences* **11:**8–13.

Abbreviated Answers to Problems

Chapter 34

1. $\Delta G^\circ = -(8.3\,\text{J/mol-K})(310\,\text{K}) \ln\left[(1 \times 10^{-3})/(1 \times 10^{-3})^4\right]$
   $\Delta G^\circ = -53.41\,\text{kJ/mol}$

2. $\Delta G^\circ = -124.6\,\text{kJ/mol}$

3. Assuming a density of 1.3 g/mL for the tubulin monomer, one can calculate that a 55-kD protein has a volume of 7.028×10^{-23} liter. Assuming the monomer is spherical, it is 2.56 nm in radius or 5.12 nm in diameter. Then 3906 monomers would be needed to span a liver cell that is 20 μm or 20,000 nm across.

4. The volume of a lymphocyte of 10 μm diameter is $5.236 \times 10^{11}\,\text{nm}^3$. The volume of an HIV particle (100 nm in diameter) is $5.236 \times 10^5\,\text{nm}^3$. If HIV particles occupy 1% of the lymphocyte volume, then a lymphocyte should contain 10,000 HIV particles.

5. To solve this problem, you must assume a certain number of individuals were already infected with HIV in 1981. If you assume about 4,000 infected people in 1981, then the current number of infected individuals (3,000,000) would be reached by 1993 if the doubling time is approximately 16 months. This increase can be best appreciated by graphing the data, as in the accompanying figure.

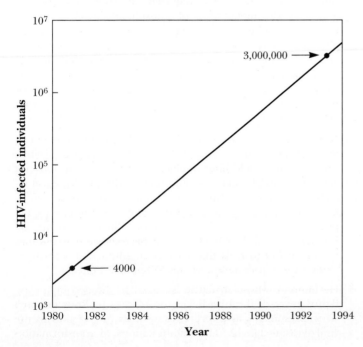

6. The human rhinovirus is made up of three different coat protein subunits that "decorate" the surface (plus a fourth subunit that is "buried" and below the protein shell of the capsid). Icosahedral viruses with three different coat protein subunits (with similar conformations) are **pseudo-equivalent**, whereas viruses with one kind of coat protein that can assume three different conformations are referred to as **quasi-equivalent.**

Chapter 35

1. $\Delta G = RT \ln ([C_2]/[C_1])$
   $\Delta G = +4.0\,\text{kJ/mol}$

2. $\Delta G = RT \ln ([C_{\text{out}}]/[C_{\text{in}}]) + Z\mathcal{F}\Delta\psi$
   $\Delta G = +4.08\,\text{kJ/mol} -2.89\,\text{kJ/mol}$
   $\Delta G = +1.19\,\text{kJ/mol}$
   The unfavorable concentration gradient thus overcomes the favorable electrical potential and the outward movement of Na^+ is not thermodynamically favored.

3. One could solve this problem by going to the trouble of plotting the data in v vs. [S], $1/v$ vs. $1/[S]$, or $[S]/v$ vs. [S] plots, but it is simpler to examine the value of $[S]/v$ at each value of [S]. The Hanes-Woolf plot makes clear that $[S]/v$ should be constant for all [S] for the case of passive diffusion. In the present case, $[S]/v$ is a constant value of 0.0588 $(\text{L/min})^{-1}$. It is thus easy to recognize that this problem describes a system that permits passive diffusion of histidine.

4. This is a two-part problem. First calculate the energy available from ATP hydrolysis under the stated conditions, then use the answer to calculate the maximal internal fructose concentration against which fructose can be transported by coupling to ATP hydrolysis. Using a value of -35.7 kJ/mol for the ΔG° of ATP and the indicated concentrations of ATP, ADP, and P_i, one finds that the ΔG for ATP hydrolysis under these conditions (and at 298K) is -40.1 kJ/mol. Putting the value of $+40.1$ kJ/mol into (Equation 35.1) and solving for C_2 yields a value for the maximum possible internal fructose concentration of 10.7 M. Thus, ATP hydrolysis could (theoretically) drive fructose transport against internal fructose concentrations up to this value. (In fact, this value is probably close to, if not in excess of, the practical limit of fructose solubility.)

5. Nigericin is a mobile carrier ionophore, whereas cecropin is a channel-forming ionophore. As noted in Chapter 9, the transport rates for mobile carrier ionophores are quite sensitive to temperature (in the vicinity of the membrane phase transition), but the transport rates for channel-forming ionophores,

by contrast, are relatively insensitive to temperature. The phase transition temperature for DPPC is 41.4°C (Table 9.2). Therefore, the transport rates for nigericin in DPPC should increase significantly between 35°C and 50°C, but the rates observed for cecropin *a* in DPPC membranes at 50°C should be about the same as they are at 35°C.

6. Since enolase is the glycolytic enzyme responsible for producing PEP, inhibition of enolase by fluoride would be expected to reduce the cytosolic level of PEP and thus inhibit the PEP-dependent PTS system.

7. Lactose transport into *E. coli* is coupled to proton uptake. FCCP and dicumarol are uncouplers and will dissipate the transmembrane proton gradient that is essential for lactose transport.

8. Each of the transport systems described can be inhibited (with varying degrees of specificity). Inhibition of the rhamnose transport system by one or more of these agents would be consistent with involvement of one of these transport systems with rhamnose transport. Thus, non-hydrolyzable ATP analogs should inhibit ATP-dependent transport systems, ouabain should specifically block Na^+ (and K^+) transport, uncouplers should inhibit proton-gradient-dependent systems, and fluoride should inhibit the PTS system (via inhibition of enolase).

Chapter 36

1. The pronghorn antelope is truly a remarkable animal, with numerous specially evolved anatomical and molecular features. These include a large windpipe (to draw in more oxygen and exhale more carbon dioxide), lungs that are three times the size of those of comparable animals (such as goats), and lung alveoli with five times the surface area, so that oxygen can diffuse more rapidly into the capillaries. The blood contains larger numbers of red blood cells and thus more hemoglobin. The skeletal and heart muscles are likewise adapted for speed and endurance. The heart is three times the size of that of comparable animals and pumps a proportionally larger volume of blood per contraction. Significantly, the muscles contain much larger numbers of energy-producing mitochondria, and the muscle fibers themselves are shorter and thus designed for faster contractions. All these characteristics enable the pronghorn antelope to run at a speed nearly twice the top speed of a thoroughbred racehorse and to sustain such speed for up to an hour.

2. Refer to Figure 36.16. Step 5, in which the myosin head conformation change occurs, is the step that should be blocked by β,γ-methylene-ATP, since hydrolysis of ATP should occur in this step and β,γ-methylene-ATP is non-hydrolyzable.

3. Phosphocreatine is synthesized from creatine (via creatine kinase) primarily in muscle mitochondria (where ATP is readily generated) and then transported to the sarcoplasm, where it can act as an ATP buffer. The creatine kinase reaction in the sarcoplasm yields the product creatine, which is transported back into the mitochondria to complete the cycle. Like many mitochondrial proteins, the expression of mitochondrial creatine kinase is directed by mitochondrial DNA, whereas sarcoplasmic creatine kinase is derived from information encoded in nuclear DNA.

4. Note in step 4 of Figure 36.16 that it is ATP that stimulates dissociation of myosin heads from the actin filaments—the dissociation of the crossbridge complex. When ATP levels decline (as happens rapidly after death), large numbers of myosin heads are unable to dissociate from actin, and the muscle becomes stiff and unable to relax.

5. The skeletal muscles of the average adult male have a cross-sectional area of approximately 55,000 cm². The gluteus maximus muscles represent approximately 300 cm² of this total. Assuming 4 kg of maximal tension per square centimeter of cross-sectional area, one calculates a total tension for the gluteus maximus of 1200 kg (as stated in the problem). The same calculation shows that the total tension that could be developed by all the muscles in the body is 22,000 kg (or 24.2 tons)!

Chapter 37

1. Polypeptide hormones constitute a larger and structurally more diverse group of hormones than either the steroid or amino-acid-derived hormones, and it thus might be concluded that the specificity of polypeptide hormone-receptor interactions, at least in certain cases, should be extremely high. The steroid hormones may act either by binding to receptors in the plasma membrane or by entering the cell and acting directly with proteins controlling gene expression, whereas polypeptides and amino-acid-derived hormones act exclusively at the membrane surface. Amino-acid-derived hormones can be rapidly interconverted in enzyme-catalyzed reactions that provide rapid responses to changing environmental stresses and conditions. See the *Student Study Guide* for additional information.

2. The cyclic nucleotides are highly specific in their action, since cyclic nucleotides play no metabolic roles in animals. Ca^{2+} ion has an advantage over many second messengers because it can be very rapidly "produced" by simple diffusion processes, with no enzymatic activity required. IP_3 and DAG, both released by the metabolism of phosphatidylinositol, form a novel pair of effectors that can act either separately or synergistically to produce a variety of physiological effects. DAG and phosphatidic acid share the unique property that they can be prepared from several different lipid precursors. Nitric oxide, a gaseous second messenger, requires no transport or translocation mechanisms and can diffuse rapidly to target cells and sites.

3. Nitric oxide functions primarily by binding to the heme prosthetic group of soluble guanylyl cyclase, activating the enzyme. An agent that could bind in place of NO—but that does not activate guanylyl cyclase—could reverse the physiological effects of nitric oxide. Interestingly, carbon monoxide, which has long been known to bind effectively to heme groups, appears to function in this way. Solomon Snyder and his colleagues at Johns Hopkins University have shown that administration of CO to cells that have been stimulated with nitric oxide causes attenuation of the NO-induced effects.

4. Herbimycin, whose structure is shown in the accompanying figure, reverses the transformation of cells by Rous sarcoma virus, presumably as a direct result of its inactivation of the viral tyrosine kinase. The manifestations of transformation

(on rat kidney cells, for example) include rounded cell morphology, increased glucose uptake and glycolytic activity, and the ability to grow without being anchored to a physical support (termed "anchorage-independent growth"). Herbimycin reverses all these phenotypic changes. On the basis of these observations, one might predict that herbimycin might also inactivate tyrosine kinases that bear homology to the viral pp60[v-src] tyrosine kinase. This inactivation has in fact been observed, and herbimycin is used as a diagnostic tool for implicating tyrosine kinases in cell-signaling pathways.

5. The identification of phosphorylated tyrosine residues on cellular proteins is difficult. Quantities of phosphorylated proteins are generally extremely small, and to distinguish tyrosine phosphorylation from serine/threonine phosphorylation is tedious and laborious. On the other hand, monoclonal antibodies that recognize phosphotyrosine groups on protein provide a sensitive means of detecting and characterizing proteins with phosphorylated tyrosines, using, for example, Western blot methodology.

6. Hormones act at extremely low concentrations, but many of the metabolic consequences of hormonal activation (release of cyclic nucleotides, Ca^{2+} ions, DAG, etc., and the subsequent alterations of metabolic pathways) occur at and involve higher concentrations of the affected molecular species. As we have seen in this chapter, most of the known hormone receptors mediate hormonal signals by activating enzymes (adenylyl cyclase, phospholipases, protein kinases, and phosphatases). One activated enzyme can produce many thousands of product molecules before it is inactivated by cellular regulation pathways.

Chapter 38

1. If side 1 is denoted as the side with 330 mM Na^+ and side 2 as the side with 70 mM Na^+, then the equilibrium potential $\Delta\psi = \psi_2 - \psi_1 = 40$ mV. Note that in any real situation there would have to be counteranions available, but this calculation considers only the sodium concentrations on either side of the membrane.

2. Use the permeability coefficients from Figure 38.3 and apply the Goldman equation to the concentrations as given. The calculated potential difference is -21 mV.

3. Vesicles with an outside diameter of 40 nm have an inside diameter of approximately 36 nm and an inside radius of 18 nm. These data correspond to a volume of 2.44×10^{-20} liter. Then 10,000 molecules/6.02×10^{23} molecules/mole = 1.66×10^{-20} mole. The concentration of acetylcholine in the vesicle is thus 1.66/2.44 M or 0.68 M.

4. Decamethonium and succinylcholine are acetylcholine analogs that bind to and activate the acetylcholine receptor. However, acetylcholinesterase shows no activity toward decamethonium, and succinylcholine is only slowly hydrolyzed by this enzyme. Thus, the effects of decamethonium are longer-lasting than those of succinylcholine. Succinylcholine is used as a muscle relaxant in certain surgeries, because it blocks the transmission of nerve impulses to muscles. Since it is slowly hydrolyzed by acetylcholinesterase and also by other cholinesterases in liver and in the blood, its effects diminish soon after drug treatment is stopped.

5. The evidence outlined in this problem points to a role for cAMP in fusion of synaptic vesicles with the presynaptic membrane and the release of neurotransmitters. GTPγS may activate an inhibitory G protein, releasing G_i(GTPγS), which inhibits adenylyl cyclase and prevents the formation of requisite cAMP.

6. Illumination of rod cells stimulates a phosphodiesterase that cleaves cGMP to GMP, thus closing plasma membrane Na^+ channels. The result is hyperpolarization of the plasma membrane—that is, the plasma membrane becomes more negative inside.

Glossary of Biochemical Terms

acrosome. The organelle that surrounds the head of a sperm and lies just inside and juxtaposed to the plasma membrane. p. 1215

action potentials (nerve impulses). Transient changes in the membrane potential that move rapidly along nerve cells. p. 1221

active transport system. A system of transporting molecules across a membrane in which the transported species moves from a low concentration to a high concentration; it requires energy input. p. 1131

alternating conformation model. A controversial model describing the mechanism of glucose transport; it states that the glucose-binding site is alternately exposed to the cytoplasmic and extracellular surfaces of the membrane. p. 1130

anion transport system. A facilitated diffusion system of the erythrocyte membrane. p. 1130

antiport. In active transport, the simultaneous transport of two different molecules in opposite directions. p. 1142

axon. A long, thin projection extending from the cell body of a neuron; its primary function is to carry nerve impulses from the cell body to the cellular extremities or termini. p. 1219

axoneme. A complex bundle of microtubule fibers that includes two central, separated microtubules surrounded by nine pairs of joined microtubules. p. 1107

bacteriorhodopsin. The light-driven H^+-pump. p. 1140

Ca^{2+}-induced Ca^{2+}-release (CICR). A process whereby Ca^{2+}, leaking through the sarcolemma Ca^{2+} channels, triggers the release of even more Ca^{2+} from the sarcoplasmic reticulum. p. 1176

calciosomes. Membrane vesicles that are similar to muscle sarcoplasmic reticulum. p. 1197

cardiac glycosides (cardiotonic steroids). Plant and animal steroids that specifically inhibit Na^+, K^+-ATPase, and ion transport. p. 1134

cell body (soma). In neurons, the part of the cell containing the nucleus and other organelles, such as the endoplasmic reticulum and mitochondria. p. 1219

chemical potential difference (ΔG). The free energy difference between side 1 and side 2 of a membrane; for an uncharged molecule, $\Delta G = G_2 - G_1 = RT \ln([C_2]/[C_1])$; for a charged molecule, $\Delta G = G_2 - G_1 = RT \ln([C_2]/[C_1]) + Z\mathscr{F}\Delta\psi$, where Z is the charge on the transported species, $\mathscr{F}$ is Faraday's constant, and $\Delta\psi$ is the electric potential difference across the membrane. p. 1126

cilia. Short, cylindrical, hairlike projections on the surfaces of the cells of many animals and lower plants; they serve to move cells or to facilitate the movement of extracellular fluid over the cell surface. p. 1106

class II major histocompatibility complex. A complex involved in the recognition of foreign proteins by the immune system. p. 1122

concentration gradient. The difference in concentrations of species on opposite sides of a membrane; given by $[C_2] - [C_1]$. p. 1126

cytoskeleton. An intracellular scaffold formed of microtubules, microfilaments, and intermediate filaments. p. 1106

dendrites. Short, highly branched structures emanating from the neuron cell body, from which they receive neural impulses and to which they transmit them. p. 1219

diffusion coefficient (D). A term tha[t] describes the rate of diffusion of a mole[cule]. p. 1126

electrical synapses. Synapses characte[r]ized by a very small gap between the pr[e]synaptic cell and the postsynaptic cel[l]. p. 1228

electrogenic transport. A transport pr[o]cess that results in a net movement [of] electric charge across the membran[e]. p. 1131

endothelins. A family of homologo[us] regulatory peptides, synthesized by ce[r]tain endothelial and epithelial cells, tha[t] act on nearby smooth muscle and co[n]nective tissue cells. p. 1243

essential light chain. In myosin, the LC[1] polypeptide chain. p. 1160

facilitated diffusion. The transport [of] substances across a membrane, as facil[i]tated by proteins. p. 1127

gap junction structures. Structures tha[t] permit the passive flow of small molecul[es] from one cell to another. p. 1148

general porins. Bacterial membran[e] proteins that form nonspecific por[es] across the outer membrane and sort mo[l]ecules according to molecular siz[e]. p. 1145

group translocation. A transport proces[s] in which a chemical modification accom[]panies the transport. p. 1143

halorhodopsin. The light-driven Cl[] pump of *Halobacterium halobium*, an a[r]chaebacterium that thrives in a high-sa[lt] medium. p. 1140

hormones. Chemical substances tha[t] serve as metabolic signals to coordina[te] the many and diverse processes tha[t] occur in different parts of an organis[m]. p. 1180

human immunodeficiency virus type [1] (HIV-1). A membrane-coated virus tha[t]

Glossary of Biochemical Terms

acrosome. The organelle that surrounds the head of a sperm and lies just inside and juxtaposed to the plasma membrane. p. 1215

action potentials (nerve impulses). Transient changes in the membrane potential that move rapidly along nerve cells. p. 1221

active transport system. A system of transporting molecules across a membrane in which the transported species moves from a low concentration to a high concentration; it requires energy input. p. 1131

alternating conformation model. A controversial model describing the mechanism of glucose transport; it states that the glucose-binding site is alternately exposed to the cytoplasmic and extracellular surfaces of the membrane. p. 1130

anion transport system. A facilitated diffusion system of the erythrocyte membrane. p. 1130

antiport. In active transport, the simultaneous transport of two different molecules in opposite directions. p. 1142

axon. A long, thin projection extending from the cell body of a neuron; its primary function is to carry nerve impulses from the cell body to the cellular extremities or termini. p. 1219

axoneme. A complex bundle of microtubule fibers that includes two central, separated microtubules surrounded by nine pairs of joined microtubules. p. 1107

bacteriorhodopsin. The light-driven H^+-pump. p. 1140

Ca^{2+}-induced Ca^{2+}-release (CICR). A process whereby Ca^{2+}, leaking through the sarcolemma Ca^{2+} channels, triggers the release of even more Ca^{2+} from the sarcoplasmic reticulum. p. 1176

calciosomes. Membrane vesicles that are similar to muscle sarcoplasmic reticulum. p. 1197

cardiac glycosides (cardiotonic steroids). Plant and animal steroids that specifically inhibit Na^+, K^+-ATPase, and ion transport. p. 1134

cell body (soma). In neurons, the part of the cell containing the nucleus and other organelles, such as the endoplasmic reticulum and mitochondria. p. 1219

chemical potential difference (ΔG). The free energy difference between side 1 and side 2 of a membrane; for an uncharged molecule, $\Delta G = G_2 - G_1 = RT \ln([C_2]/[C_1])$; for a charged molecule, $\Delta G = G_2 - G_1 = RT \ln([C_2]/[C_1]) + Z\mathscr{F}\Delta\psi$, where Z is the charge on the transported species, $\mathscr{F}$ is Faraday's constant, and $\Delta\psi$ is the electric potential difference across the membrane. p. 1126

cilia. Short, cylindrical, hairlike projections on the surfaces of the cells of many animals and lower plants; they serve to move cells or to facilitate the movement of extracellular fluid over the cell surface. p. 1106

class II major histocompatibility complex. A complex involved in the recognition of foreign proteins by the immune system. p. 1122

concentration gradient. The difference in concentrations of species on opposite sides of a membrane; given by $[C_2] - [C_1]$. p. 1126

cytoskeleton. An intracellular scaffold formed of microtubules, microfilaments, and intermediate filaments. p. 1106

dendrites. Short, highly branched structures emanating from the neuron cell body, from which they receive neural impulses and to which they transmit them. p. 1219

diffusion coefficient (D). A term that describes the rate of diffusion of a molecule. p. 1126

electrical synapses. Synapses characterized by a very small gap between the presynaptic cell and the postsynaptic cell. p. 1228

electrogenic transport. A transport process that results in a net movement of electric charge across the membrane. p. 1131

endothelins. A family of homologous regulatory peptides, synthesized by certain endothelial and epithelial cells, that act on nearby smooth muscle and connective tissue cells. p. 1243

essential light chain. In myosin, the LC1 polypeptide chain. p. 1160

facilitated diffusion. The transport of substances across a membrane, as facilitated by proteins. p. 1127

gap junction structures. Structures that permit the passive flow of small molecules from one cell to another. p. 1148

general porins. Bacterial membrane proteins that form nonspecific pores across the outer membrane and sort molecules according to molecular size. p. 1145

group translocation. A transport process in which a chemical modification accompanies the transport. p. 1143

halorhodopsin. The light-driven Cl^- pump of *Halobacterium halobium*, an archaebacterium that thrives in a high-salt medium. p. 1140

hormones. Chemical substances that serve as metabolic signals to coordinate the many and diverse processes that occur in different parts of an organism. p. 1180

human immunodeficiency virus type 1 (HIV-1). A membrane-coated virus that is the causative agent in acquired immune deficiency syndrome (AIDS). p. 1121

immunoglobulin gene superfamily. A large class of molecules that generally serve in cellular recognition processes. p. 1122

internal coupling. The association of terminal cisternae with t-tubules, forming a dyad junction. p. 1176

interneurons. Neurons that simply pass signals from one neuron to another. p. 1219

ionophore antibiotics. Small molecule toxins produced by microorganisms; they facilitate ion transport across membranes. p. 1149

mechanochemical coupling. Motion coupled with a chemical reaction. p. 1107

microtubules. Hollow, cylindrical structures formed from the protein tubulin. p. 1105

mobile carriers. Molecules that form complexes with particular ions and diffuse freely across a lipid membrane. p. 1149

motor neurons. Neurons that pass signals from other neurons to muscle cells, thereby inducing muscle movement (motor activity). p. 1220

multidrug resistance (MDR). The simultaneous resistance of cells to an original drug and a wide spectrum of drugs with little structural or functional similarity to the original drug. p. 1139

muscle fibers. The long, multinucleate cells of skeletal muscles. p. 1157

muscle types. Four different kinds of muscle found in animals: skeletal muscle, cardiac muscle, smooth muscle, and myoepithelial cells. p. 1157

myasthenia gravis. An autoimmune disease that causes muscle weakness and eventually paralysis. p. 1234

myocytes. Cells containing a single nucleus. p. 1157

myofibrils. Linear arrays of cylindrical sarcomeres, the basic structural units of muscle contraction. p. 1157

neuroglia (glial cells). Cells that serve to protect and support nerve functions. p. 1219

neurons. Cells that carry out the reception and transmission of nerve impulses. p. 1219

neurotransmitter (chemical synapse). A chemical substance that binds to receptors on the postsynaptic cell and initiates a new action potential. p. 1228

nodes of Ranvier. Gaps between Schwann cells along an axon's length. p. 1219

olfactory epithelium. A layer of receptor cells (and supporting cells) surrounding the air spaces of the nasal passages. p. 1251

oncogenes. Genes implicated in tumor formation; they code for proteins that are capable of stimulating cell growth and division. p. 1189

osteoblast. A bone cell that synthesizes the collagen fibrils that form the bone's organic matrix. p. 1138

osteoclast. A multinucleate cell that breaks down bone during normal bone remodeling. p. 1137

partition coefficient (K). The ratio of the solubility of a molecule in a hydrophobic solvent (similar to the membrane core) to solubility in water. p. 1126

passive diffusion. A transport process in which species move across a membrane in the thermodynamically favored direction without the help of any specific transport system or molecule. p. 1126

peptide hormones. Hormones derived from amino acids; they regulate processes in body tissues, including the release of other hormones. p. 1181

peripheral coupling. The association of terminal cisternae with the sarcolemma membrane in a dyad structure. p. 1176

permeability coefficient (P). A term that describes how readily a molecule leaves the water solvent and crosses the hydrophobic barrier presented by a membrane. It is usually expressed in cm/sec. p. 1126

plus end. The end of a microtubule where growth occurs; the other end is called the *minus end*. p. 1106

pore (channel). A molecule that adopts a fixed orientation in a membrane, creating a hole that permits the transmembrane movement of ions. p. 1149

porins (peptidoglycan-associated proteins; matrix protein). Membrane proteins through which small polar molecules and ions can diffuse. p. 1145

postsynaptic cell. The synaptic cell that receives the neural signal. p. 1228

preprohormones Peptide hormones synthesized from mRNA as inactive precursors; they become activated by proteolysis. p. 1206

presynaptic cell. The synaptic cell that delivers the neural signal. p. 1228

proto-helices. Short, helical segments of just over two turns. p. 1110

proto-oncogenes. Normal, noncancerous genes involved in cell growth regulation. p. 1189

proton-motive force. The energy stored in a proton concentration gradient across a membrane. p. 1142

quasi-equivalent symmetry. A term used to describe a virus structure composed of coat proteins with identical amino acid sequences but different conformations. p. 1115

rate of flow or flux (J_c). The transport rate per unit area of an uncharged molecule C across a membrane; given by $J_c = -P(C_2 - C_1)$. p. 1126

regulatory light chain. In myosin, the LC2 light peptide chain. p. 1160

retrovirus. A virus that carries RNA as its genetic material. p. 1121

sarcolemma. The muscle fiber plasma membrane. p. 1157

sarcoplasmic reticulum (SR). A specialized endoplasmic reticulum that covers the sarcomere. p. 1157

Schwann cells. Glial cells that envelop and surround neurons to form a protective sheath. p. 1219

second messengers. Cyclic nucleotides, Ca^{2+} ions, or other substances that activate or inhibit enzymes or cascades of enzymes in very specific ways. p. 1181

secondary active transport. Transport that depends on an ion gradient for its energy input. p. 1131

self-assembling complexes. Macromolecular polymeric structures formed spontaneously from certain biological molecules. p. 1102

sensory neurons. Neurons that acquire sensory signals, either directly or indirectly from receptor cells, and pass this information along to interneurons or to motor neurons. p. 1219

sensory transduction. The process of sensing the environment and turning the acquired information into nerve impulses. p. 1243

signal transduction unit. A hormone, receptor, and associated transduction system. p. 1187

sliding filament model. A model describing how the thick myosin filaments slide or walk along the thin actin filaments when muscle fibers contract. p. 1168

specific porins. Membrane proteins that contain binding sites for particular substrates. p. 1145

steroid hormones. Hormones derived from cholesterol that regulate metabolism, salt and water balances, inflammatory processes, and sexual function. p. 1181

striated muscle. Skeletal and cardiac muscle, which contain microscopic alternating light and dark bands. p. 1157

symport. In active transport, the simultaneous transport of two different molecules in the same direction. p. 1142

synapse (synaptic cleft). The space between a synaptic knob and a dendrite ending. p. 1219

synaptic knobs (synaptic bulbs). Small structures on the distal end of the axon of a nerve cell. p. 1219

T4 lymphocytes. A class of white blood cells, also known as helper T cells. p. 1121

transverse tubules (t-tubules). Extensions of the sarcolemma that enable the sarcolemmal membrane to contact the ends of each myofibril in the muscle fiber. p. 1157

treadmilling. A process in which tubulin monomers are added to the plus end of a microtubule at about the same rate at which monomers are removed from the minus end. p. 1106

triad. The structure at the end of each sarcomere, which consists of a t-tubule and two apposed terminal cisternae. p. 1158

tumor promoters. Agents that do not themselves cause tumorigenesis but that potentiate the effects of carcinogens. p. 1202

tumor suppressor genes. Genes that code for proteins whose normal function is to turn off cell growth. p. 1190

viral transforming proteins. Proteins produced by oncogenic viruses that transform animal cells to a cancerous state. p. 1204

voltage-gated ion channels. Transaxonal membrane proteins that act as voltage-regulated pores to allow the passage of ions such as Na^+, K^+, and Ca^{2+}. p. 1221

Index

Photo and Illustration Credits

Part V Opener: courtesy of David Barford, Oxford University

Chapter 34 opener: Dallas Skyline at Night; Werner Krutein/Liaison International; **34.1** (influenza): KG Murti/Visuals Unlimited; (rotavirus): CNRI/Photo Researchers, Inc.; (polyoma) Science VA/Visuals Unlimited; (rat sperm) David Phillips/Visuals Unlimited; (connective tissue) Fawcett and Heuser/Photo Researchers Inc.; (ciliate) Eric Grave/Phototake; (nasal ciliate): Veronika Burmeister/Visuals Unlimited; (microtubule) KG Murti/Visuals Unlimited; (soil bacteria) Dr. Tony Brain/Custom Medical Stock; **34.2:** Adapted from a drawing by Gerald Stubbs; **34.8:** Adapted from a drawing by Ronald Vale; **34.4:** M. Schliwa/Visuals Unlimited; **34.9:** Dr. Jeremy Burgess/Photo Researchers, Inc.; **34.19:** J.C. Revy/Phototake

Chapter 35 opener: "Drawbridge at Arles with a Group of Washerwomen" (1888) by Vincent van Gogh; Rikjsmuseum Kroller-Muller/photo by Erich Lessing/Art Resource; **p. 1135:** (viceroy butterfly) Patti Murray/Animals, Animals; (monarch butterfly) E. R. Degginger/Animals/Animals; (foxglove) Zig Leszczynski/Animals, Animals; (lily of the valley) John Bova/Photo Researchers, Inc.; (oleander) Arthur Hill/Visuals Unlimited; (milkweed) L. West/Photo Researchers, Inc.; **35.39:** (cecropia moth) Greg Neise/Visuals Unlimited; (cecropia moth caterpillar) Patti Murray/Animals, Animals

Chapter 36 opener: Michelangelo's "David"; the Firenze Academia/Art Resource; **36.3:** courtesy of S. Fleischer and M. Inui; **36.4** and **36.6** courtesy of Hugh Huxley, Brandeis University; **36.7 (a and b)** courtesy of Linda Rost and David DeRosier, Brandeis University; **(c)** courtesy of George Phillips, Rice University; **36.8:** courtesy of Henry Slayter, Harvard Medical School; **36.9 (ribbon image)** and **36.16 (inset):** courtesy of Ivan Rayment and Hazel M. Holden, University of Wisconsin, Madison; **36.20, 36.21,** and **36.24:** courtesy of S. Fleischer

Chapter 37 opener: Drawing of Embryo in the Uterus by Leonardo da Vinci; Superstock Int.; **37.37:** courtesy of David Barford, Oxford University

Chapter 38 opener: Spiders and Snakes from Albert Seba's "Locupletissimi Rerum Naturalium" (ca 1750) hand-coloured engraving; The Bridgeman Art Library; **p. 1224: (left)** Zig Leszczynski/Animals, Animals; **(right)** Tim Rock/Animals, Animals **38.11:** courtesy of M. Montal; **38.31:** Prof. P. Motta/Dept. of Anatomy/University "La Sapienza," Rome/Science Photo Library; **38.43:** courtesy of Randall Reed, Johns Hopkins University.

A List of Common Abbreviations Used by Biochemists

A	adenine (or the amino acid alanine)
Ab	antibody
Ag	antigen
Ac-CoA	acetyl-coenzyme A
ACh	acetylcholine
ACP	acyl carrier protein
ADH	alcohol dehydrogenase
ADP	adenosine diphosphate
AIDS	acquired immunodeficiency syndrome
AMP	adenosine monophosphate
ALA	δ-aminolevulinic acid
Ala	alanine
Arg	arginine
Asn	asparagine
Asp	aspartate
ATCase	aspartate transcarbamoylase
atm	atmosphere
ATP	adenosine triphosphate
BChl	bacteriochlorophyll
bp	base pair
BPG	bisphosphoglycerate
BPheo	bacteriopheophytin
C	cytosine (or the amino acid cysteine)
cal	calorie
CaM	calmodulin
cAMP	cyclic 3',5'-adenosine monophosphate
CAP	catabolite activator protein
cDNA	complementary DNA
CDP	cytidine diphosphate
CDR	complementarity-determining region
Chl	chlorophyll
CM	carboxymethyl
CMP	cytidine monophosphate
CoA or CoASH	coenzyme A
CoQ	coenzyme Q
cpm	counts per minute
CTP	cytidine triphosphate
Cys	cysteine
D	dalton (or the amino acid aspartate)
d	deoxy
dd	dideoxy
DAG	diacylglycerol
DEAE	diethylaminoethyl
DHAP	dihydroxyacetone phosphate
DHF	dihydrofolate
DHFR	dihydrofolate reductase
DNP	dinitrophenol
Dopa	dihydroxyphenylalanine
DNA	deoxyribonucleic acid
E	glutamate
$\mathscr{E}$	reduction potential
E4P	erythrose-4-phosphate
EF	elongation factor
EGF	epidermal growth factor
EPR	electron paramagnetic resonance
ER	endoplasmic reticulum
F	phenylalanine
$\mathscr{F}$	Faraday's constant
F_{AB}	antibody molecule fragment that binds antigen
FAD	flavin adenine dinucleotide
$FADH_2$	reduced flavin adenine dinucleotide
FBP	fructose-1,6-bisphosphate
FBPase	fructose bisphosphatase
Fd	ferredoxin
fMet	N-formyl-methionine

FMN	flavin mononucleotide
F1P	fructose-1-phosphate
F6P	fructose-6-phosphate
G	guanine or Gibbs free energy (or the amino acid glycine)
GABA	γ-aminobutyric acid
Gal	galactose
GDP	guanosine diphosphate
GLC	gas-liquid chromatography
Glc	glucose
Gln	glutamine
Glu	glutamate
Gly	glycine
GMP	guanosine monophosphate
G1P	glucose-1-phosphate
G3P	glyceraldehyde-3-phosphate
G6P	glucose-6-phosphate
G6PD	glucose-6-phosphate dehydrogenase
GS	glutamine synthetase
GSH	glutathione (reduced glutathione)
GSSG	glutathione disulfide (oxidized glutathione)
GTP	guanosine triphosphate
H	histidine
h	hour
h	Planck's constant
Hb	hemoglobin
HDL	high-density lipoprotein
HGPRT	hypoxanthine-guanine phosphoribosyltransferase
HIV	human immunodeficiency virus
HMG-CoA	hydroxymethylglutaryl-coenzyme A
hnRNA	heterogeneous nuclear RNA
HPLC	high-pressure (or high-performance) liquid chromatography
HX	hypoxanthine
Hyl	hydroxylysine
Hyp	hydroxyproline
I	inosine (or the amino acid isoleucine)
IDL	intermediate density lipoprotein
IF	initiation factor
IgG	immunoglobulin G
Ile	isoleucine
IMP	inosine monophosphate
IP_3	inositol-1, 4, 5-trisphosphate
IPTG	isopropylthiogalactoside
IR	infrared
ITP	inosine triphosphate
J	joule
K	lysine
k	kilo (10^3)
K_m	Michaelis constant
kb	kilobases
kD	kilodaltons
L	leucine
L	liter
LDH	lactate dehydrogenase
LDL	low density lipoprotein
Leu	leucine
Lys	lysine
M	methionine
M	molar
m	milli (10^{-3})
Met	methionine
mL	milliliter
mm	millimeter

Man	mannose
Mb	myoglobin
mol	mole
mRNA	messenger RNA
mV	millivolt
N	asparagine
N	Avogadro's number
n	nano (10^{-9})
NAD$^+$	nicotinamide adenine dinucleotide
NADH	reduced nicotinamide adenine dinucleotide
NADP$^+$	nicotinamide adenine dinucleotide phosphate
NADPH	reduced nicotinamide adenine dinucleotide phosphate
NAG	N-acetylglucosamine
NAM	N-acetylmuramic acid
NANA	N-acetylneuraminic acid
nm	nanometer
NMR	nuclear magnetic resonance
O	orotidine
P	phosphate (or the amino acid proline)
p	pico (10^{-12})
P$_i$	inorganic phosphate
PAGE	polyacrylamide gel electrophoresis
PBG	porphobilinogen
PC	plastocyanin (or the phospholipid phosphatidylcholine)
PCR	polymerase chain reaction
PE	phosphatidylethanolamine
PEP	phosphoenolpyruvate
PEPCK	PEP carboxykinase
PFK	phosphofructokinase
PG	prostaglandin
2PG	2-phosphoglycerate
3PG	3-phosphoglycerate
PDGF	platelet-derived growth factor
PGI	phosphoglucoisomerase
PGK	phosphoglycerate kinase
PGM	phosphoglucomutase
Phe	phenylalanine
Pheo	pheophytin
PIP$_2$	phosphatidylinositol-4,5-bisphosphate
PK	pyruvate kinase
PKU	phenylketonuria
PLP	pyridoxal-5-phosphate
Pol	polymerase
PP$_i$	pyrophosphate ion
PQ	plastoquinone
Pro	proline
PRPP	5-phosphoribosyl-1-pyrophosphate
PS	photosystem (or the phospholipid phosphatidylserine)
PTH	phenylthiohydandoin
Q	coenzyme Q, ubiquinone (or the amino acid glutamine)
QH$_2$	reduced coenzyme Q (ubiquinol)
R	gas constant (or the amino acid arginine)
r	revolution
RER	rough endoplasmic reticulum
RF	release factor
RFLP	restriction fragment length polymorphism
RNA	ribonucleic acid
rRNA	ribosomal RNA
R5P	ribose-5-phosphate
Ru1,5P	ribulose-1,5-bisphosphate
Ru5P	ribulose-5-phosphate
S	Svedberg unit (or the amino acid serine)
s	second
s	sedimentation coefficient
SAM	S-adenosylmethionine
SDS	sodium dodecylsulfate
Ser	serine
snafu	situation normal, all fouled up
snRNA	small nuclear RNA
snRNP	small nuclear ribonucleoprotein
SRP	signal recognition particle
T	thymine or absolute temperature (or the amino acid threonine)
TCA	tricarboxylic acid cycle
TDP	thymidine diphosphate
THF	tetrahydrofolate
Thr	threonine
TMP	thymidine monophosphate
TMV	tobacco mosaic virus
TPP	thiamine pyrophosphate
Trp	tryptophan
TTP	thymidine triphosphate
Tyr	tyrosine
U	uracil
UDP	uridine diphosphate
UDPG	UDP-glucose
UMP	uridine monophosphate
UTP	uridine triphosphate
UV	ultraviolet
V	valine, volt
Val	valine
V_{max}	maximal velocity
VLDL	very low density lipoprotein
W	tryptophan
X	xanthine
XMP	xanthine monophosphate
Xu5P	xylulose-5-phosphate
Y	tyrosine
yr	year
YAC	yeast artificial chromosome